Praise

'This book is the ideal companion for anyone pondering a move into contracting, providing thorough coverage of essential knowledge. I wish I had a time machine to bring it back to the beginning of my contracting journey.'
— **Marti Floyd**, DevOps Engineer, MGBOps Ltd

'Confident Contractor is a game-changer for anyone stepping into the world of contracting. It's like having a mentor by your side, offering clear guidance and practical tips to navigate the complexities of the industry. With its straightforward approach and real-life examples, this book has given me the confidence to step out of my comfort zone to become self-employed. A must-read for aspiring contractors!'
— **Elizabeth State,** Social Media Marketer

'This book completely demystifies the world of contracting, arming you with everything needed to take that daunting leap and be successful thereafter. An indispensable toolkit which I will be turning to repeatedly for years to come on my contracting journey.'
— **Kieran Stockdale**, DevOps Engineer, Equal Experts

Confident
Contractor

Thriving in IT Contracting and beyond

Neil Millard

R^ethink

*This book is dedicated to my wife,
Wendy, who gets me and tries to keep
me on track with my crazy ideas.*

Contents

Foreword

Since the Covid-19 pandemic, it is more challenging than ever to be self-employed in the UK.

While the number of self-employed workers steadily increased for twenty years to peak at 5 million in 2019, it had fallen to closer to 4 million in 2022, according to the Office for National Statistics.[1] A study by the Centre for Economic Performance at the London School of Economics and Political Science paints

---

1   Office for National Statistics, 'Understanding changes in self-employment in the UK: January 2019 to March 2022' (ONS, 6 July 2022), https://bit.ly/ONS_2022, accessed April 2024

a bleak picture of stagnant incomes, rising costs and risk to mental health amongst the self-employed.[2]

While the idea of being your own boss and having a higher earning potential is still as appealing as ever – especially for highly skilled technologists in the IT sector – it is trickier to make it work than it has been in the past. There is less room for trial and error, less room for amateurism. Now more than ever, it is important to know what you are doing and to be aware of the ins and outs of contracting, so that you can be a confident contractor.

Over the last fifteen years building my consulting business, I have met literally thousands of seasoned IT contractors as well as would-be first-time contractors in the UK and across the world. In this time, I have been struck by how differently people approach contracting. While many truly understand that it is about running a business first and foremost, there are still some who behave as if they were

---

still employees. They do not understand their new responsibilities, hoping to get all the benefits without fundamentally changing their perspective. In the current economic downturn, this difference in approach and understanding is what will make all the difference between a rewarding contracting career and a disappointing one.

Neil Millard was one of the most experienced contractors that I met during this time. Neil possesses a deep understanding of both the risks and rewards of contracting and has accumulated a wealth of knowledge over the years that he is sharing in this book.

The best thing about this book is its simplicity – it is free of jargon and is particularly well suited for people who have absolutely no experience in contracting. Even so, it will also provide a precious resource for people who might have got into contracting already but have never taken the time to reflect on what they need to do to be successful after their first contract.

The book covers all the basics so that you can get started safely, and demystifies the more arcane topics like cashflow and IR35. Beyond that, it will provide a practical framework that encompasses all aspects of contracting so that you can build a sound mental model of what it takes to succeed in this field before you jump into it. The knowledge contained in this book will resolve the common misunderstandings and false assumptions people have about contracting, so that you can decide for yourself if it's for you and where to start.

Remember: being your own boss comes with responsibilities, and you can't just wing it. At the same time, self-employment is a great alternative to traditional employment. You can have more agency, more control over your future and in the mutually beneficial relationships you build with customers and team members. Armed with this book as your guide, you will surely thrive – happy contracting.

Thomas Granier
CEO and co-founder of Equal Experts
www.linkedin.com/in/thomasgranier
www.equalexperts.com

Introduction

If you've picked up this book, then at some point – maybe right now – you have thought about plying your IT skills through contracting.

You've felt a spark that fills you with joy, been swept away by dreams of 'what if?' What if you could be more, write books, entertain a crowd with your presentations, provide the best solutions to problems and not be confined by the boundaries of someone else's business?

You want to run your own IT business, but are not sure what it involves or how to go about it. You want to become a confident contractor.

I know these feelings. Having been made redundant twice in five years, I realised that a regular salaried job was not for me and I looked for a different way.

Maybe you have already taken the leap but are using the services of an umbrella company. You are providing awesome solutions for your customer, charging a fair rate, but the umbrella company sends only a small amount to your bank account, compared to your rate stated on the contract. Where does the rest go? Is there an alternative where you can claim your full value and get the most out of your work?

Being a contractor, freelancer or, as one of my associates described it, 'working independently through my own company,' means not having a boss, a salary or performance reviews. It means no line management – of yourself or others. You are hired to perform or deliver a specific skill or service, nothing more. As an IT professional contractor for over twenty years, I can teach you how to start, run and profit from your IT contracts via your own limited company. My experience spans three decades that have seen me through being a field engineer, a manager and team leader, through to the professional IT contractor I am today.

I've written this book to provide you with the blueprint I never had. In it, I share my experiences building multiple businesses, securing various contract renewals, renegotiations and terminations. My successes and my mistakes have helped to guide me, and now they can guide you through the forest of complexity that is running a business.

Becoming an IT contractor is daunting and is not a decision anyone takes lightly. There are lots of variables to consider and decisions to make. Through reading this book you will learn what it is to be a confident contractor and how to excel in your new role as a director of a limited company. You will know what to do and when do to it. I will give you the answers to questions that you might not have thought to ask.

This book will be your guide through all the stages of running a successful IT contractor business:

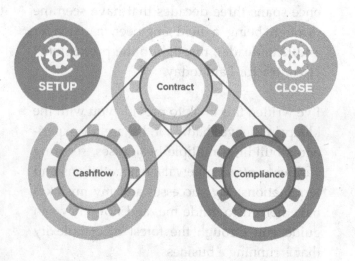

In the first chapters, I will go through the start-up phases and tasks, beginning with naming your company. Then, we dive into the three Cs to become a confident contractor: Contracts, Cashflow and Compliance.

We will explore how to secure and begin a contract, learn what the requirements of running a smooth operation are and get clear on the implications of IR35. This will enable you to focus on delivering your best self and building business systems that support you. I will explain what should be in an IT contract, where to find contracts and how to prepare for interviews and negotiations. I will then give you my model for success once you've secured a contract, with schedules and reminders for the various reporting requirements you will be responsible for.

Your business will die without good cashflow. I will teach you my easy-to-follow system to ensure the right amount of cash is flowing into and out of your business to ensure things are ticking over and you achieve the lifestyle you want.

Next, we'll talk about compliance and tax, and the regular recording and reporting of monthly, quarterly and yearly operations that must be completed and is easy to do once you have the right systems in place. I'll walk you through what to do with the money you will receive, with a blueprint to follow so that you and your family can live without worry. Whether you want to buy a house, have more holidays or save for your best retirement, you will have choices.

Finally, we will talk about what happens when a contract comes to an end, what provisions you should have in place and what you can do when you no longer want or need the company.

PART ONE
FOUNDATIONS OF IT CONTRACTING

Like any property, project or software, for successful IT contracting you need strong foundations that you can rely on to serve you over the long term.

We'll start this book by learning what those foundations are and how to build them, and going through what you need to do to prepare for your first contract.

1
Introduction To IT Contracting

The IT contracting sector is highly dynamic: the technology and tools we use in IT come and go, and successful IT contracting requires you to embrace the change and adapting your services with the ebbs and flows of the market. Flexibility and adaptability are paramount.

As independent contractors, as well as the technical skills needed to get the job done, we must also have the ability to run a small business and manage our own careers. This requires a unique skill set that must be constantly honed and updated.

To succeed in IT contracting, you must recognise and understand these dynamics and be able to identify the shifting opportunities and challenges as they arise. The contracts that will define your professional journey and determine your trajectory can vary from short-term, intensive projects to long-term, immersive engagements. Each contract can demand upgrades to and diversification of your skill set, and you cannot shy away from this. Understanding the fluidity of this sector is the first step towards building a successful career.

The IT contracting landscape

The employment dynamics in the IT sector, once characterised by traditional structures and contracts, have changed in recent years. The gig economy has emerged and expanded, and the demand for specialised skills has led to IT contracting becoming a prominent work model. No longer are IT professionals mere cogs in the corporate machinery; as contractors, they are versatile experts tasked with diverse challenges in an ever-evolving technological environment.

This dynamism and fluidity means that IT contracting rests on a fundamental virtue: flexibility. Free from the rigid structures of traditional employment models, IT contractors have the autonomy to cherry-pick the projects that resonate with their skills and passions. This freedom extends far beyond the mere selection of projects; you can choose your work environment, schedule and the clients you want to collaborate with. You can opt to specialise in a particular area, branch out in a different direction or take a broader approach. The choice is yours.

But flexibility alone is not enough. Adaptability is the trait that transforms a good contractor into an exceptional one. As we can all observe, technology is not static; by nature, it is constantly evolving, with new innovations and products emerging all the time, and others falling out of use. As industry trends shift and client expectations change with them, a confident contractor will need to embrace a mindset of continuous learning and adaptation.

The adaptability required goes beyond mastering a particular programming language or a specific project management methodology. It demands a commitment to staying abreast

of technological advancements, being percep-
tive to industry shifts and proactively learning
new skills. Information technology is a living
thing, and those who can swiftly adjust their
skills, tools and approaches will find themselves in high demand.

The adaptability required in IT contracting
extends to interpersonal skills. Clients too are
not monolithic entities. Understanding their
changing needs, anticipating their expectations and aligning your approach accordingly
is crucial, as is the ability to communicate
complex technical concepts to non-technical
stakeholders or pivot project strategies based
on client requests.

In a field where the only constant is change, IT
contractors often find themselves on the cutting
edge of innovation. They become the pioneers
who experiment with emerging technologies,
refine their skills in real-world scenarios and
bring a wealth of experiential knowledge to
the organisations they engage with. This posi-
tions them not only as problem solvers but as
thought leaders in the tech space. It's a hugely
exciting effort to be involved in.

Yet with the myriad opportunities that IT contracting presents, there are also challenges. Chief among these is that the contracts can be unpredictable, requiring contractors to be financially savvy and maintain a robust professional network. A successful IT contractor must not only deliver exceptional technical expertise but also master the business side of contracting, ensuring a steady stream of projects and sustainable income.

The best time to enter the IT contracting market is today. Start looking now: gauge where the market is, identify the tools and skills that are 'trendy'. You cannot time the market but be prepared and ready to launch when the opportunity arises. So that you are prepared, let's take a closer look at what benefits IT contracting offers and what it takes to be successful in this environment today.

The advantages of IT contracting

In the contemporary professional world, IT contracting is unique in the level of flexibility and autonomy it permits. While traditional employment structures are characterised by rigid schedules and corporate bureaucracy, IT contracting allows professionals to chart their own course. This flexibility is the big advantage of contracting, empowering individuals to determine when and where they work, allowing them to tailor their schedules to optimise productivity and achieve their desired work–life balance.

Autonomy, mastery and purpose

In his book *Drive*, Daniel Pink argues that intrinsic human motivation can be divided into autonomy, mastery and purpose.[3] All three of these can be found in IT contracting. Autonomy allows you to be creative in designing a solution to the problem your customers are facing and build something that will make a difference in the world and have purpose. This independence is important in a few areas. You can work free from the shackles and limitations of your client environment; this separation is important when considering IR35 status, which we'll discuss later.

Mastery in the context of IT contracting can be found in the freedom to select projects that

---

3   D Pink, *Drive: The surprising truth about what motivates us* (Canongate Books, 2018)

align with the skills and interests that you want to master. When these are aligned with your purpose, this fosters a profound sense of professional fulfilment.

Purpose can be found in choosing your own professional path, which not only enhances job satisfaction but also enables contractors to continuously evolve and grow professionally.

Money

Financial rewards are another compelling benefit of IT contracting for those who can be strategic. Unlike traditional employment contracts, where compensation is usually predetermined, IT contractors can negotiate competitive rates based on their skills and the market demand for their expertise.

This financial control and higher earning potential comes with the responsibility of managing your own taxes and finances – this is a critical task and demands financial literacy in the contracting sphere.

Opportunities

The gig economy, of which IT contracting is a significant part, has become a major force reshaping traditional modes of employment. Companies are increasingly turning to contractors for specialised skills, allowing them to access a diverse pool of talent without the long-term commitment or boundaries of traditional employment.

As a contractor you can be removed from the team quicker, so you pose less risk to the hiring manager. Without the lengthy recruiting and onboarding process for employees, you can meet on Wednesday and start work on Monday. For IT contractors, this translates into a wealth of opportunities to collaborate with various organisations, work on exciting projects and expand your professional networks.

The IT contractor's skill set

To be a successful IT consultant requires a specific skill set. One of the skills you will need to have early on is **selling yourself**. In traditional employment, you only need to get one contract, for a job that will have been advertised.

In contracting, you will need to win (and renew) lots of contracts. In the beginning, finding and engaging with clients is difficult. The interview process is short and usually technical, followed by a 'do we like you' exercise or chat. It's all about answering the main two questions that clients have:

1. Can you deliver the technical skills required by the project?

2. Will you upset anyone, or are you aligned with the values of the client?

Compared to the process often used to hire for a salaried job, which may consist of multiple technical tests, various HR processes and tick box exercises, winning a contract is usually just a case of one or two meetings before moving on to contract negotiation.

The negotiation at the beginning of a relationship is quite different to that of renewing an ongoing contract. **Strategic negotiation** has emerged as a crucial skill for IT contractors seeking to maximise their financial gains. You must be able to effectively communicate the value you bring to a project and negotiate fair compensation in order to build a sustainable and lucrative contracting career.

The ability to **articulate the unique skills and expertise** that make you an invaluable asset distinguishes you as a successful negotiator in this sector. This skill not only ensures fair compensation but also establishes a foundation for long-term, mutually beneficial relationships with clients.

A state of **perpetual learning** not only keeps you professionally relevant but also enhances your marketability. The freedom to choose projects aligned with personal interests further encourages this continuous learning journey, driving ongoing professional growth and development.

Contracting also nurtures a **mindset of entrepreneurship**. Contractors are not just workers; they are business owners responsible for managing every aspect of their professional lives. This requires and indeed cultivates a spirit of innovation, resourcefulness and resilience. IT contractors are not only constantly adapting to change but actively seeking new opportunities and creating value in an evolving marketplace.

Your **professional profile** and **networking abilities** will play a pivotal role in your success as an IT contractor. Building and maintaining a robust professional network opens doors to new projects, collaborations and potential clients. The freedom to choose projects that speak to your personal interests further facilitates the forging of meaningful professional connections. Networking is not just about expanding your circle; it's about fostering relationships that will lead to long-term partnerships and opportunities.

Risks and challenges

Of course, if IT contracting was entirely risk-free and offered only advantages, everybody would be doing it. When embarking on a contracting career, there are several key risks and challenges to be aware of.

Cashflow

Cashflow is a critical part of any business. When embarking on the journey of IT contracting, you must be prepared for the startup costs and keep one eye on what's coming in and going out.

Like any new business, you will incur costs and even when you're not being paid, you still have to live. Let me explain. If you are currently in a permanent position with a salary, you will probably be provided with a desk, a laptop and a monthly payday. On day one of contracting, you will need a laptop and a place to work, but there will be a lag until your next payday. Let's assume you've got paid on the last day of your job. Your contract, once you have one, could pay monthly too, but you'll have to work for a whole month before you can send the first invoice, which might not be paid until four weeks later. That's eight weeks from your last payday as an employee. This means that you'll need to have a little stash of cash on hand to last you until then, and to pay for that new laptop and other expenses. This will be your buffer.

You may well have a budget in your personal life, where you record the income and outgoings of your household. Doing so will help you understand the cashflow of your business. We'll go into more detail about cashflow management later in the book. But with a personal budget in place and a cash buffer in your business, you will be all set financially to leap into contracting.

Insecurity

Work insecurity is an inevitable part of IT contracting. Unlike the (perceived) stability of a permanent role, contracting involves defined durations and project-based work, intensifying the need for contractors to have resilience and a proactive mindset. Contractors must accept and be able to deal with the reality that their engagement is temporary. Even if a lengthy contract is signed, there will be a break clause. Day One is not the only time you will need that stash of cash.

To navigate this insecurity successfully, a fundamental shift in mindset is required. You need to not only accept change but embrace it as an opportunity for growth. This insecurity is one of the reasons that the day rate for a contractor is higher compared to an employee. You should use this to build a buffer in your business; this will enable you to pay off those early expenses and build some stability into your cashflow. One of the biggest mistakes made by new contractors is removing too much money from the business too soon, reducing the buffer.

Dealing with insecurity is often about perspective. While in some ways contracting increases insecurity, in other ways it removes it. During my early years in the IT industry, I felt insecure in my permanent position, with two of my early jobs ending in redundancy. Having a contract removes the uncertainty by putting a date on it. When you know it's coming, you can be prepared.

As a contractor, you can turn job insecurity into an advantage by actively cultivating a diverse skill set that gives you broad appeal and increases the pool contracts you can pitch for. In the fast-paced world of IT, technologies evolve rapidly and staying relevant is crucial. Diversifying your skills not only makes you more marketable as a contractor but also increases your adaptability. By constantly updating your technical proficiencies and staying informed about emerging industry trends, you can position yourself as an asset, ready to take on a wide spectrum of challenges and step in at relatively short notice compared to companies having to recruit and onboard a new employee.

Networking also plays a key role in mitigating job insecurity. Building and maintaining connections within professional communities opens doors to new opportunities and insights. Networking not only aids in finding new contracts but also enables you to build a support system to help guide you in weathering the ups and downs of the contracting journey, by providing referrals and advice. Engaging with fellow professionals, attending industry events and participating in online forums can significantly contribute to your ability to ride out ebbs and flows of the market and the uncertainties inherent in IT contracting.

It is through networking that most of my contracts have found me. Helping a customer on a project in Bristol in 2005 led to a personal recommendation when that client wanted my help again in 2012. In the beginning though, growing your network is about graft. This means ensuring your online profiles and website are up to date, updating your CV and speaking to as many recruitment agencies as possible.

Taking a wider view, the economy of course plays a part in determining the demand for and supply of work, which affects day rates as well as the general availability of projects requiring our expert services. Like many things, the economy moves in a cycle of ups and downs, and your profile and network will be particularly important during down phases.

In general, managing your pipeline of IT job contracts requires a proactive approach that prioritises adaptability, continuous learning and a robust professional network. By embracing change as an inherent part of the contracting journey, you can use job insecurity as a catalyst for professional development and grow a successful business.

Compliance

For IT contractors, the other main risk is that of tax and IR35. An understanding of the tax regulations that apply to you is a crucial aspect of ensuring both legal compliance and financial optimisation. Among the myriad regulations that apply to you you'll need to get to grips with, IR35 stands out as a common challenge that demands careful consideration and understanding.

IR35, or 'off-payroll working' regulations, were designed to prevent tax avoidance by individuals working as contractors who should be classified as employees for tax purposes. The key determination made by IR35 is whether a contractor is a genuine contractor or a 'disguised employee'.

Changes in UK tax law mean that the effect of IR35 on your tax bill is now greatly reduced from what it had been. Tax reforms have increased the tax liability of operating off-payroll (outside IR35) to bring near equal tax liability, whether you are on or off payroll. Yet misclassification, such as taxable income being based on the day rate, rather than what is paid out by the company, has consequences. The risk can be mitigated by understanding your contract and its implications and careful examination of working arrangements.

Navigating IR35 requires a proactive approach. Contractors should seek professional advice, engage in regular assessments of their working arrangements and stay informed about any changes in legislation.

Your contractor identity

Having a diverse array of skills and a broad knowledge base can keep you in demand, but only if people know about them. To ensure people are aware of you and what you can offer, you'll need to build a professional identity as a contractor. This will be shaped not just by the projects you undertake but by the values you bring to your work. Crafting your contractor identity involves defining your professional ethos, the skills you want to be known for and the impact you aim to make in the industry.

Key to your identity will be your unique selling proposition (USP). Understanding what this is for you sets the foundation for a successful contracting career. Whether you specialise in a niche technology or possess a rare skill set, your contractor identity should distinguish you in a competitive market.

It helps also to have an online identity that sets out your skills and project background. You are, after all, always looking for that next project, and having an online presence and portfolio enables you to be seen. Having potential clients approach you gives you the upper hand, enabling you to pick and choose. As a serious business you will want to have a website and email address that aligns with your business name. There are some setup costs associated with setting up a website and there are ongoing costs from year to year to run the domain, website and email. All of this is well worth it. Once you have a website, set up relevant online profiles to complement and link back to it.

These are the basics, but you could then consider starting a blog where you can talk and be seen, which might lead to new opportunities. Do you want to help others grow? Consulting, mentoring and training could be part of your business if you want it to. Giving talks, publishing papers and attending conferences can further build your profile. You might even go as far as writing a book.

Personal and professional goals

Considering your personal and professional goals as a contractor extends beyond financial aspirations. It's about achieving a harmonious work–life balance that supports your overall wellbeing. While the flexibility of contracting offers opportunities for a balanced lifestyle, it also requires intentional effort in time management and self-care.

As a contractor, you must define your personal and professional goals holistically, thinking about not only any financial milestones you want to reach but also what kind of life you want. Balancing the demands of contracts with finding personal fulfilment is an ongoing process that, done right, will allow you to enjoy long-term satisfaction in the contracting profession.

Your financial goals should consider how you want to budget your money and your time, giving you enough of both to pursue your interests. Done right, contracting will give you more money to manage, and more time to 'spend' on what you want to do.

Summary

To succeed as a contractor, you need to understand the transformative shift that is taking place in employment dynamics and the opportunities this creates for you if you can be adaptable and proactive.

There are many advantages of IT contracting, including the **unmatched flexibility and autonomy** that you can enjoy, as well as **higher earning potential**. When you're ready for a new role, the interview process is usually a lot shorter than it is for salaried employment and once you have templates and systems in place, a contract can be signed in minutes.

But along with the benefits, there are risks and challenges, like the inherent **job insecurity** (you can be removed from a team as fast as you can be added to it) and the need to understand and comply with the intricate IR35 regulation. **Cashflow management** is an unavoidable part of running a business and can create risk if ignored. To reduce the risks and stress of starting out, you should create a personal budget and build a buffer in your business.

To fully reap the rewards of contracting and overcome or mitigate the potential risks, there are certain key skills and attributes that will set you up for success as an IT contractor, key among which are **resilience** and an **attitude of continuous growth**.

You will need to craft a unique **contractor identity** based on your USP, values and aspirations. This should allow you to achieve a harmonious work–life balance that is aligned with your personal and professional goals.

2
Setting Up Your Company

As a contractor, you have the freedom to work on the projects and with the clients that you choose. You also have the potential to earn more than if you were employed by a company. But it also means running your own business, which comes with its own challenges, as we touched on in the last chapter, not least of which are managing your finances and taxes.

You will also need to decide what kind of company you want, which relates to how much freedom and control you want in managing your contracts. The first decision is whether to start out working under an umbrella company, where contracts are managed for you, or whether you want to be in full control.

Umbrella company or limited company?

When operating as an IT contractor, you are working in the world of Business to Business (B2B) contracts. You have two main options for managing your work and finances: setting up as a limited company yourself or using an umbrella organisation.

An umbrella organisation is basically an employment agency, meaning they take care of the paperwork and admin and ensure compliance with tax and employment laws. The umbrella organisation will sign the contract, send invoices and pay you as an employee (also taking care of PAYE tax). This greatly simplifies the complexity of being a contractor but comes at the cost of less efficient tax management and reduced control over your cashflow. It also comes with a fee, of course.

The world of umbrella companies is vast and unregulated, and there is a huge range of choice. Recently there have been reports of some less scrupulous organisations 'forgetting' to pay the tax owed and leaving their customers with a large and unexpected bill. My advice is to be cautious and choose carefully if you decide to have an umbrella organisation assist you.

When running your own limited company, you gain options that can lead to greater tax efficiency, as you can take advantage of Corporation Tax and dividend rates (which are lower than rates for Income Tax). You also have more control over your finances and how you operate your business: you sign your own contracts, raise the invoices yourself, file annual accounts and tax returns, choose how much to pay yourself and when, and pay all taxes yourself direct to the government. This does come with some drawbacks: more paperwork and admin.

I said you usually have the choice. If your contract is 'inside' IR35, then you will be forced to use one from a list of 'approved' umbrellas. We'll learn more about IR35 later in the book. This isn't a choice you have to stick with. You can change from one model to the other when a new contract is about to be signed. I started by using the umbrella model for my first contract and then started a limited company when I became more confident.

While an umbrella company might help you get started, one of the most important things that you can do on your journey to becoming a confident contractor is to set up your own limited company. This will give you several advantages, including:

- **Limited liability:** As a director of a limited company, you have limited liability. This means that your personal assets are protected if the company goes bankrupt.

- **Tax benefits:** There are a number of tax benefits available to limited companies. For example, you can pay yourself dividends, which are taxed at a lower rate than Income Tax.

- **Professionalism:** Setting up a limited company signals that you are serious about your business and that you are committed to providing a professional service to your clients.

The rest of this chapter will walk you through the process of setting up your own limited company, step by step, from choosing a company name to planning for the future. In it, we will discuss the different legal structures and business registration requirements, setting up business bank accounts and working out tax implications and liabilities.

Choosing your company name

Your company name is a big part of your brand. It should resonate with your customers and reflect your business's essence. You need to decide this early on, as you'll need a name to register your company. When choosing a company name, there are a few things to keep in mind: it should be unique, memorable, relevant to your business and easy to pronounce.

Your company name must not already be in use by another similar company, so you will need to check if it has already been registered. For the UK there are three registrars: the Registrar of Companies for England and Wales, the Registrar of Companies for Scotland and the Registrar of Companies for Northern Ireland. The rules are broadly similar regardless of which Registrar of Companies you are operating under.

To check whether your desired company name is available, you can check the Companies House website. Beware of other websites that provide this information for a fee. There are (currently) no fees for accessing data about companies, only for filing information. The official site is part of the gov.uk domain.

Your company name will be central to all you do

As a contracting vehicle, the name might not be hugely important, which gives you options as simple as using your own name. But if your intention is to establish a brand, your company name should be relevant to your business. This will help your customers to easily understand what your business does and what kind of products and/or services you offer.

It is useful for people you may be working with (eg a bank manager, accountant, tax adviser, as well as clients) if your company name is easy to pronounce and remember. You will definitely want your customers to be able to remember your company name so that they can mention it to others. Another consideration when determining a business name to use, is the availability of a good domain for your website address, which should be as close to your business name as possible.

Below are some ideas you can use to come up with a business name:

- **Alliteration:** Alliteration is the repetition of consonant sounds at the beginning of words, for example, 'Creative Cloud' or 'Range Rover'.

- **Puns:** A pun is a play on words that uses two or more words with similar sounds but different meanings, for example 'Yelp' (a play on the word 'help') and 'Reddit' (which twists the phrase 'read it').

- **Foreign words:** Using a foreign word in your company name can make it stand out as unique and appear more sophisticated. For example, 'Nike' is Greek for 'victory' and 'Volvo' is Latin for 'I roll'.

- **Acronyms:** Acronyms can be a good way to create a short and memorable company name. For example, 'IBM' (International Business Machines) and 'NASA' (National Aeronautics and Space Administration).

- **Compound words:** A compound word is two or more words that are combined to form a new word, for example, 'Facebook' and 'YouTube'.

- **Descriptive words:** Words that describe your company's products or services can be a good way to choose a company name. For example, 'Microsoft' (software for microcomputers) or 'Cloudflare' (cloud-based web security services).

- **Geographic locations:** Using a geographic location in your company name can make it more memorable and relevant to your target audience. For example, 'Oxford Pizza Kitchen' or 'London Eye'.

- **Historical figures or references:** Using a historical figure or reference in your company name can make it more unique and memorable. For example, 'Tesla' (ie Nikola Tesla) or 'Apple' (referring to Isaac Newton).

- **Personal names:** Using your own name or the names of other people in your company (if relevant) helps to personalise your brand. For example, 'Dell' (founded by Michael Dell) or 'Hewlett-Packard' (Bill Hewlett and Dave Packard).

Choosing a company name is an important decision, but it doesn't have to be overwhelming. The most crucial thing is that you don't take too long about it – decide on a unique and effective name that you like and that will serve you well in the long term, so that you can get on with registering and running your business.

Legal structures and business registration

Once you have a name for your limited company, it's time to register it. But when you do so, you'll need to indicate what kind of legal structure your business will have. This is up to you. The options are:

1. **Sole trader:** You and your business are legally one and the same. You have unlimited liability and are taxed as an individual.

2. **Partnership:** Where two or more partners work together. Also unlimited liability and taxed as individuals. There are two limited liability options: Limited Liability Partnership (LLP) and Limited Partnership (one or more partners have unlimited liability).

3. **Private limited company ('Ltd'):** This is the most common type of limited company and is suitable for most businesses, including sole contractors. It is limited liability and gives options on how to distribute profits in a tax efficient manner.

4. **Public limited company ('Plc'):** This type of limited company is larger and more complex than a private limited company. It is more suitable for large businesses that need to borrow public money and that are listed on the stock exchange.

The most common legal structure for IT contractors is a limited company. A limited company is a separate legal entity from its owners, so all the company's assets and liabilities are separate from the owner's personal assets and liabilities. This can be a significant advantage for contractors, as it protects your home, for example, from being used to recover any debts or pay for any liabilities that your company may incur.

If you decide to be a sole trader, you will need to register as self-employed with His Majesty's Revenue and Customs (HMRC). To register a limited company, you will need to complete the following steps:

1. Know your company name (discussed above).

2. Decide on your company address. This is the address where Companies House will send you official correspondence.

3. Choose the company directors. Every limited company must have at least one director, who is responsible for making decisions and managing the company.

4. Decide on the share structure and shareholders. The share structure of a limited company determines how ownership of the company is divided. The shareholders are the owners of the company. This is you and anyone else you wish to share the profits with via dividends.

You will also need two documents that describe how the company will be run:

1. A Memorandum of Association: This is a legal statement signed off by the initial shareholders or guarantors agreeing to form the company.

2. Articles of Association: These are written rules about running the company agreed by the shareholders or guarantors, directors and the company secretary.

The Memorandum of Association will be created automatically during the registration process. It would be unusual to write your own Articles of Association. Typically, you would use the standard articles, known as 'model articles'.

You will also need to identify people with significant control (PSC) over your company. This is usually just yourself.

Just one more thing: your SIC (Standard Industrial Classification) code. This is a code that describes what your company does, from an agreed list of classifications. Companies House uses a condensed list available on their website, but the full list is available from the Office of National Statistics.

Your business bank account

Once your company is registered, you will need to open a business bank account. This is a different bank account from your personal bank account. It is important to keep your business and personal finances separate, as this will make it easier to keep track of your business expenses and income.

When opening a business bank account, you'll want to compare the different accounts that are available and choose one that meets your needs. You should consider the following factors:

- Any fees associated with the account

- The interest rate offered on the account

- The features and services that are included with the account

- The customer service reputation of the bank

As IT contractors, we just need a few, pretty uncomplicated features. For example, we do not need to handle cash, so counter services will rarely be required. At the basic level, we need to be able to easily receive and send money electronically. In the UK, most of the 'challenger' banks can do this, and they usually have a simple online or in-app process for opening an account.

To open the account, you will need some details about the company and personal ID for yourself as the controller of the bank account. If you already have a personal account with the same bank, this can speed things up a little.

If you intend to keep profits in the business (instead of withdrawing as dividends, for example), a savings or investment account for your business would also be useful so that you can get a return on those funds too.

VAT registration

Value-added tax (VAT) is a consumption tax that is applied to most goods and services sold in the UK. It is a tax that is paid by the consumer, but it is collected by businesses. Businesses that are registered for VAT must charge VAT on their sales and then pay it to HMRC.[4]

Most businesses that reach the turnover threshold in a twelve-month period must register for VAT. The threshold is set by budget announcements and as of the 2024/25 tax year, it is £90,000 (which equates to a day rate above £375 based on twenty billable days per month). The only exceptions are businesses that exclusively sell exempt goods or services (examples include education, insurance and membership subscriptions). As an IT contractor, you will

---

4   Gov.uk, Register for VAT, www.gov.uk/register-for-vat

almost certainly need to register for VAT if you reach the threshold.

You can also choose to register for VAT voluntarily when your turnover is less than £85,000. This can be beneficial if you want to reclaim VAT on your expenses or want to be able to charge VAT to your customers.

To register for VAT, you need to complete an online application form on the HMRC website. You will need to provide some information about your business, such as your name, address and turnover. You will also need to choose a VAT accounting period. This is usually every three months (quarterly), starting in the month you apply. Once you have submitted your application, HMRC will review it and send you a VAT registration number. When you receive that number, you must start charging VAT from that date. You will need to charge VAT on most goods and services that you sell to UK customers. As an IT contractor, you will almost certainly need to charge VAT. The standard rate of VAT, as of 2024, is 20%.

You must then keep records (invoices and receipts) of all your sales and purchases so that you can complete a VAT return every quarter or year, depending on the VAT accounting period you have chosen. You can reclaim any VAT paid on goods and services that you purchase for your business on your VAT return.

Summary

When setting up your company, the first step is **choosing a name** and **building your brand**. The name should be memorable, relevant and easy to pronounce, and not in use by anyone else. Brainstorm ideas using techniques like alliteration, puns or foreign words.

There are then several steps to legally set up your business. You will first need to choose a suitable **legal structure**, based on your needs. If you decide on a limited company, you'll need to register it with Companies House, provide certain information and prepare some legal documents. You then need to open a dedicated **business bank account** and **register for VAT** if your turnover exceeds the threshold. As well as charging VAT to your customers you can also claim it back on eligible business purchases via your VAT return.

Remember, you are not alone. There are many other contractors who have successfully set up their own limited companies and, if you follow my earlier advice and grow your professional network, you will have plenty of people to ask for help if you get stuck. By following the steps outlined in this chapter and seeking professional advice when needed, you can build a strong foundation for your business.

3
Negotiating And Managing Contracts

With the setup done, you are ready to find and sign a contract. Your network of friends and recruiters will be the first to tell you about opportunities and roles. In addition, there are online portals that advertise contract roles alongside permanent jobs.

You can also take a proactive approach by reaching out directly to companies or organisations where you would like to work. Research target companies, identify decision-makers or hiring managers, and craft personalised messages expressing your interest in potential contract opportunities. Highlight your relevant skills, experience and value to capture their attention.

Contracts come in various forms and lengths, depending on their purpose, but most contracts share some common elements. Understanding your contract is vital, and in this chapter, we'll dive into its key sections and talk through the common terms used so that you can speak the language and know what each paragraph is doing. I will then share with you some of my strategies for negotiating on contract terms and tips on how to manage contracts once you're in them, as well as what to do when they end.

While this is a guide, it is not a comprehensive legal aid and any contract should be reviewed by a legal professional to ensure that it is fair and correct. In addition, the contract should be verified by IR35 specialists, if you want the contract to be treated as off-payroll.

IR35

Before we get into the detail about contracts, we need to discuss IR35, an employment regulation designed to ensure that contractors pay roughly equivalent Income Tax and National

Insurance (NI) to employees, and that employment status cannot be manipulated or misrepresented to avoid tax.[5]

Most contracts you encounter will fall inside or outside of IR35 regulation. As of April 2021, the responsibility for determining if a contract is inside or outside IR35 lies with the end client (or employment agency) who pays your limited company. This can cause lots of confusion, as some organisations may err on the side of caution and state that the contract is inside IR35, when case law may suggest that it is outside. Legislation prevents any blanket determination, so it comes down to a contract-by-contract basis.

The laws about IR35 are so fluffy that even HMRC find it difficult to win court cases related to them.[6] As a result, there is a lot of case law. I hope the rules will be simplified soon, as it is a major pain point in the industry. The information below is a broad summary, but laws and regulations can of course change,

5   Gov.uk, 'Understanding off-payroll working (IR35)', www.
    gov.uk/guidance/understanding-off-payroll-working-
    ir35, accessed April 2024

6   N Nordone, 'Qdos shuts down three-year-long £100,000
    IR35 investigation' (Qdos, 21 March 2023), www.
    qdoscontractor.com/ir35/qdos-shuts-down-ir35-
    investigation, accessed April 2024

so always seek professional advice to ensure you remain compliant.

Unsurprisingly, for a 'contractor' the contract is critical. The terms and conditions outlined within it should align with the characteristics of genuine self-employment rather than an employer–employee relationship. This refers to things like the level of control the client exerts over the contractor, the right to substitution and the degree of financial risk assumed by the contractor.

Understanding working arrangements is equally vital. If the contractor operates under the client's direction and control in a manner akin to an employee, they may be deemed a 'disguised employee' by IR35, attracting higher tax liabilities and potential penalties. A genuine contractor should retain a high degree of autonomy, providing their services as an independent entity.

I will attempt to simplify the rules here, and there is an official tool created by HMRC, check employment status for tax (CEST), available on their website (linked in the Further Resources section). You can also get a contract checked by a specialist firm that will provide a Status Determination Statement (SDS) to the end client.

The elements to assess are:

- **Control:** This concerns the degree to which the end client has control over the contractor.

- **Substitution:** The ability for the contractor to assign work to a third party.

- **Mutuality of obligations:** Is the contractor obligated to accept every assignment provided and is it compulsory for the end client to offer work?

- **Equipment:** Does the contractor or the employer provide equipment?

- **Financial risk:** The degree of financial risk the contractor takes on.

- **Running a business:** Is the contractor running a business in their own right? Having a website and a business email address would indicate so.

For example, if the end client has control, provides equipment and there is a mutual obligation to give and accept work, then the contract is inside IR35.

If you are using your own equipment, can delegate or reject work and there is no mutual obligation to provide or take work, the contract is outside the scope of IR35 and you are self-employed. A contract determined outside IR35 means you are free to run your company as you see fit, with all the privileges and responsibilities that come with this.

If you have a contract inside IR35, you will be treated as an employee for tax purposes. This is essentially like using an umbrella company model and you will be paying Income Tax on most of the contract earnings. This is very often unfair, as it would be unusual for the contract to include billable time for holidays and sick pay, which a direct employee would be entitled to. This is the kind of treatment that the regulation is intended to prevent, as it could be considered exploitation by the contracting client and a way to reduce their overall costs.

I would highly recommend becoming a member of the Association of Independent Professionals and the Self-Employed (IPSE) as they offer extensive protection and campaign for contractor rights, especially in relation to IR35.

Understanding and reviewing contracts

A contract is an agreement between two or more parties. It puts into writing several key factors about the agreement between the contracting parties, as well as answering specific questions that may arise during the contract period.

Below are some common sections you will see in a contract:

- **Parties involved:** Contracts identify the parties involved, such as the client company and your company as the service provider.

- **Terms and conditions:** This section outlines the specific terms and conditions of the agreement. It may include the scope of work, the services to be provided, responsibilities, timelines and payment details. This is sometimes in a separate section or document called the Statement of Work.

- **Key project information:** This is a summary of the agreement, which will include the project duration with start and end dates, location, any specific requirements like professional indemnity (PI) insurance, consideration and termination periods.

- **Location:** This is where the services are expected to be performed or provided – it could be all on-site, hybrid or fully remote.

- **Consideration:** Consideration refers to what each party will receive in exchange for their obligations. For IT professionals, this could be a project, day or hourly rate.

- **Confidentiality and non-disclosure:** Many IT contracts contain clauses related to confidentiality and non-disclosure, particularly if sensitive information or proprietary data is involved.

- **Data protection:** Considerations regarding how sensitive information will be used and stored.

- **Intellectual property rights:** This concerns anything the contractor creates, or knowledge used in the delivery of services. This must clarify what software you will be creating or using from elsewhere. If in doubt, seek legal advice.

- **IR35:** There should be specific clauses related to the IR35 legislation to ensure that you and your customer are covered.

- **Liability:** This is what you could be sued for in the event of negligence or wrongdoing. Most unfortunate events can be insured for via PI insurance.

- **Indemnities:** What the contract says you are *not* liable for.

- **Termination:** Contracts often stipulate the conditions under which the agreement can be terminated, including notice periods and penalties.

- **Dispute resolution:** This section outlines the process for resolving disputes, which could include mediation, arbitration or legal action.

- **Governing law:** This should specify the jurisdiction and laws that will govern the agreement in case of disputes.

In addition, there will be a schedule with the definitions and interpretations of specific terms in the contract. For example, if a day rate is involved, what is defined as a 'day'?

A contract visualised

If in any doubt about any of these elements, it's wise to consult with a lawyer who can provide specific guidance and ensure your rights and interests are protected.

Example Statement of Work

The main contract will cover the bulk of the regular clauses laying out the responsibilities and limitations to both of the parties to the contract, but some of the contract will have variables that are detailed in a Statement of Work (SoW) or as a schedule within the contract documents. This further clarifies the specific services, fees and schedule for the project. See the example below:

Agreement between:

Your Customer Limited (registered in England and Wales under number 3456789), whose registered office is at 12 High Street, Anytown, AN1 2AB.

Your IT Contract Company Limited (registered in England and Wales under number 10234567), whose registered office is at 15 Applecart Street, Othertown, OT1 1AR.

Details

Start Date: 29 February 2024

Term: 12 months until 28 February 2025

Location: Customer Site, Newtown, NT1 8AT

Client: Customer name

Subcontractor: Your name

Services

The subcontractor will deliver the following services for the Term at the Location:

- DevOps
- Programming
- Technical Design

Fees

The subcontractor shall charge for the services provided on the basis of time spent as follows:

In any week, the subcontractor shall charge £500 per day, for a minimum period of at least 7 hours per day.

Negotiation strategies

Once you have the contract agreement in principle, you may have room to negotiate. Remember, a contract is a two-way street and both parties should be getting something they want.

Contract negotiations generally occur before a contract is signed, and at a renewal of the contract. They can occasionally happen during the contract term, should something change that either party wants to renegotiate on, although this is rare. I have negotiated at all these points.

The main points for negotiations are usually the items in the SoW: the start date, duration, services, location and rate can all be agreed upon as part of the negotiations.

To help during the negotiations, you should get to know your customer by creating a positive atmosphere and demonstrating active listening. Your goal is to foster collaboration through respectful communication and a focus on mutual understanding. This makes it easier to reach an agreement. Being patient and persistent will help, especially if you can be flexible and adaptable in the pursuit of your goals. Staying calm and professional, by controlling your emotions even if the other party becomes agitated, will help you reach an agreement.

You should be clear and concise, asking open-ended questions to gather information about your customer's needs, concerns and potential concessions. If the customer wants something from you (a concession), such as an earlier start date or longer contract length, then you can negotiate and ask for something in return.

For example, during contract negotiations, a past client of mine wanted to ensure that I could deliver services beyond the current contract end date. We scheduled a meeting to discuss my terms for renewal with a longer end date in mind. For my side, it was fair to ask for a higher consideration and, in return, the client would secure my services for a longer period. Win-win.

Be prepared to make concessions but do so strategically, focusing on value creation for both parties. Being prepared to walk away, if necessary, gives you great strength during negotiations. If you have a backup plan, such as other offers of contracts, you can walk away from a contract if it doesn't meet your requirements and saves you from a bad deal.

Negotiating day rates

Contract extension is a great time to negotiate your day rate as you have already demonstrated your value to the client, which puts you in a stronger position. I have secured some great contract renewals this way, especially when the market demand is high.

You may have to demonstrate why your day rate should increase. You can cite new skills, demonstrable returns the customer has enjoyed because of your efforts, or simply provide evidence that the market rate has changed.

Beware, though, that the market can fall as well as rise. For example, during the Covid-19 pandemic, due to the overall reduction in output of goods and services, many clients had to make the hard decision of reducing project costs, either through reduction of billable hours and people, or reduction in day rates.

Since then, remote working has provided a great benefit in the form of more flexible and inclusive working conditions and less commuting (which is more planet-friendly), as well as various other advantages. This has also increased the supply of professionals in some areas of the UK, and the customer may make a concession that does not require your presence in the office. Prior to this pandemic-triggered shift, there was only a finite number of IT professionals within an hour's commute to London. If remote working is available, the commute isn't an issue and the pool of IT professionals gets much bigger. As a result of this greater supply, day rates become more competitive. So you can see how the market fluctuates, and the change isn't always on the side of the IT contractor.

Outside factors can also affect your starting position of day rates and contract length. The state of the market and the number of available contractors will determine a 'market rate' for your services. Greater supply means lower prices, and greater demand means higher prices. This is true of any market.

You can boost your market rate by building your contractor profile, as mentioned earlier, for example by having skills and experience within specific areas of IT. Technology does not stay still so you must always have one eye on the market to see which skills are most in demand. In my contracting journey, I have specialised in WinTel, PHP, DevOps, platforms and data engineering. Having a wide range of skills makes it easier for you to appeal to more of the market as tides shift.

Ultimately negotiation is a process of give and take to reach a mutual agreement on terms that satisfies both parties. Remember, consulting a lawyer can be invaluable if you have any doubts or questions about your contract.

Managing the contract

Once the contract is signed and you're engaged in the project, you should stay alert to further opportunities. As a bare minimum, you should be regularly checking that what you are delivering is in line with the customer's expectations. If you have any questions or concerns, raise them promptly. Is anything hindering your ability to deliver what you agreed? Can you exceed expectations given a little more support in specific areas? Does the client want to retain your services at the end of the agreed term?

Sometimes a client will agree a short-term contract with you in order for both sides to build trust and understanding. If you like the project, it is in your best interest to continue delivering great value for the client, as when they are ready to discuss further engagement, you will be able to demonstrate what you have delivered to secure better terms in the renewal process.

Of course, the opposite can be true. Maybe the contract/project has not panned out in the way you were expecting. It's not unheard of to terminate a contract earlier than the agreement but be aware that there will be consequences to doing so. These should have been made clear in the contract, but sometimes the customer might surprise you.

For example, I signed a three-month contract where, verbally, the location of work was agreed to be in the south of England. Imagine my surprise when I was asked to report to a customer site in Northumberland. That is quite a commute. The customer paid for fuel and overnight accommodation, but that didn't consider the time taken to travel and that I had commitments in the southwest. Needless to say, I requested contract termination, as was my right under the contract, and I invoiced as agreed. But now that the contract had been terminated, the payment terms were not clear. I was paid – but four months later.

Building a relationship with the customer is key to project delivery and helps establish rapport. Even if the project is not going well, keeping open communications with the customer can prevent escalation. Share updates, address any delays and proactively discuss issues, and both parties will feel confident and relaxed.

Some projects involve a lot of processes, such as change control, documentation and reporting. You should adhere to these wherever possible, though it might be possible to challenge some of them. Remember, providing a quality service throughout the contract will leave you in good standing if you want to renew your contract.

The management of the contract day-to-day will essentially be you delivering the service you agreed; on a week-to-week or month-to-month basis (depending on what has been agreed in the contract), it involves submitting an invoice for that service, usually accompanied by an authorised timesheet; alternatively, if your contract sets out a project rate, or some other unit, then you would invoice against that instead.

Invoicing will not happen automatically. You will have to raise the invoices yourself, or at least task someone or something else with doing this for you. If you forget to raise the invoice, no one is going to remind you. Set up alerts in your calendar to ensure you are invoicing at regular intervals, otherwise you may find your business bank account a bit short on funds when you come to pay yourself.

Invoicing and payment process

At some point, as agreed in the contract, you will invoice your customer. This is usually monthly, or sometimes weekly. The invoice must contain specific features to be considered a valid invoice:

- Your company name with registered address

- Customer name with their registered address

- Invoice number and date of invoice

- VAT number (if registered)

- Product service quantity and value

- VAT amount (if applicable)

- Subtotals and totals

Some optional things can be included like a company logo, terms of payment and due date, bank account details for payment and purchase order (PO) number or reference. Most accounting software will provide a template that you can use, and it should look something like the invoice below.

Invoice

Your company Name

Address line 1
Address line 2
City, Postcode

To

Customer Company
Customer Name
Address
City, Postcode

Invoice Date:	04/06/2024
Invoice Number:	1234
Client Reference:	
DUE DATE:	04/06/2024

Description	Qty	Unit	Unit Price	VAT %	VAT	Total
Service/Project	20	day	£ 500	20%	£ 2000.00	£ 10,000.00

Sub Total	£ 10,000.00
Total VAT	£ 2000.00
Total amount due	£ 12,000.00

Registered Address	Payment Details	
Address line 1	Bank Name	Barclays Bank PLC
Address line 2	Sort-code	60-00-00
City, Postcode	Account No.	12345678

A PO number gives your customer a reference to tie the invoice back to the order or instruction they gave. In IT contracting, though, this instruction is the signed contract and is usually authorised by way of a timesheet. If your contract states an hourly or day rate, then someone will be able to authorise a timesheet, which, when returned to the customer with the invoice, confirms that the service has been delivered in line with the contract.

Late payment

At some point in your contracting journey, you may be unfortunate enough to not receive a payment of an invoice by the date it is due. In the first instance, you should contact your customer's finance department, or the relevant individual, as there may be a reasonable explanation for the late payment. Your customer might just need reminding or perhaps there is a missing timesheet or other information they are waiting for.

If there is no reasonable explanation for the late payment, then this would be a dispute. A good contract should cover what happens in this situation, and provide a clear guide about what to do next.

Staying in contact with your customer is the best way to resolve any dispute, but you may have to withdraw your services after a certain period of time if it looks like you are not going to receive payment.

Contract end

As you get to the end of the contract, the obvious question is: what happens next?

If you have built and maintained a good relationship with your customer, it should be obvious if they want to renew your contract. If it isn't, you can assume a renewal is not forthcoming and should start thinking about where the next contract will be coming from.

Consider whether you want to take some time to rest, train and recuperate. As a general rule of thumb, you should start looking for the next contract about six to eight weeks before you want to start. Interviews can take a little while to set up, but the whole process from looking to starting can be as short as one week. In my experience, depending on whether the client is backfilling (replacing another contractor who's leaving) or it's a new project, the process can take anywhere from one week to eight weeks.

Summary

Contracts are essential documents for IT free-lancers, **outlining the terms of your engagement** with clients. IR35 is an important regulation to get to grips with, and the contract terms are key to identifying whether the work falls within or outside of the scope of **IR35**. It can be complicated but with the correct guardrails and insurance in place, will not cause you any undue concerns.

The anatomy of an IT contract typically comprises sections on the involved parties, the scope of work and responsibilities, key project information, confidentiality and data protection clauses, intellectual property rights, and legal considerations. Schedules or SoWs often provide specific definitions and interpretations of terms used within the contract.

Negotiating your contract is key to ensuring a mutually beneficial agreement. Remember, it's a two-way street, so focus on building rapport and reaching win-win situations. Typical negotiation points include the start date, duration, services offered, location and your day rate. Effective negotiation strategies involve being clear and concise in your communication, demonstrating the value you bring, understanding the market rate for your skills and being prepared to walk away if the terms aren't working for you.

Managing your contract effectively is crucial for a successful contractor experience. Stay alert to opportunities and potential issues within the project, promptly raise any concerns you might have and consistently deliver great value to your client. Building a strong relationship with the client is key, and proactively suggesting improvements to contract processes can strengthen your standing. Remember to invoice regularly to manage your cashflow and start looking for your next contract before the current one ends to ensure a smooth transition.

By understanding, negotiating and effectively managing your contracts, you can ensure a successful and rewarding IT contracting career.

PART TWO
COMPLIANCE AND CASHFLOW

Now that you are up and running, we can turn our focus to operating your business. Invoicing, book-keeping and paying yourself will fall into a regular cadence, as will reporting to the authorities, filing on time and ensuring the business stays compliant with regulations.

4
Compliance And Legal Matters

Beyond your contractual obligations and HMRC's requirements, a clear understanding of legal and compliance issues is essential. This chapter delves into data protection and the importance of safeguarding personal identifiable information (PII), explores intellectual property ownership challenges, outlines information you must disclose to Companies House and guides you through essential business insurance considerations.

Data protection

In the IT field, which by nature deals with vast volumes of information (in the form of data), data protection has become a big concern for both businesses and individuals. In the United Kingdom, the General Data Protection Regulation (GDPR) is a comprehensive framework that governs the processing and storage of personal data.

IT contractors must fully understand their responsibilities regarding data protection. This includes knowing the lawful basis for processing data, obtaining explicit consent when necessary and implementing robust security measures to safeguard sensitive information. Failure to comply with GDPR can result in severe penalties, including hefty fines, making it imperative for IT contractors to prioritise data protection in their contractual agreements.

Your responsibilities will be defined either in the contract or in a separate data protection policy released by the client concerned. If you are working with data, then data protection will be a top consideration when working with and developing systems that hold or use that data. If any data leaks occur, you will be expected to contribute to the postmortem and offer support and suggestions to prevent further incidents.

Should your business hold any PII, then you will need to register with the Information Commissioner's Office (ICO) and specify who will be responsible for data protection in your limited company. If your business only deals with B2B contracts, then it will usually be exempt from the requirement to register.

Intellectual property rights

Intellectual property (IP) is a crucial aspect of IT, as your role often involves the creation of software, applications or other innovative solutions for your clients. It is essential for contractors to clearly define the ownership and usage rights of IP in their contracts. For example, contracts should explicitly outline whether the IP developed during the project will belong to the client or you, the contractor. Additionally, contractors should consider specifying the scope of use for the IP, ensuring that both parties have a clear understanding of how the developed assets can be utilised.

In some cases, contractors may also need to address pre-existing IP, ensuring that they have the right to use any third-party assets in their work. By carefully delineating IP rights in the contract, both parties can avoid disputes and legal complications down the line. The same applies to software licences from third parties, ensuring that the customer has the current and future right to use that software and/or IP.

Likewise, if your limited company is creating IP under its own identity, this should be registered in some way, as you may need to defend against someone else using it without permission or licence. If that intellectual property or licence is of significant value, I would recommend you seek the advice of a specialist IP lawyer.

Companies House

As a director of a limited company, you will be responsible for reporting certain information about your company to Companies House.[7] Any change to the business, for example in its structure or operations, must be assessed and

---

7   Gov.uk, 'Running a limited company: Your responsibilities', www.gov.uk/running-a-limited-company/company-changes-you-must-report, accessed April 2024

reported in accordance with the Companies Act.[8] Changes that need to be reported include:

- Registered office
- Business contact
- Directors' personal details (such as change of address)
- PSC and their details
- Company secretary

As well as reporting these changes to Companies House, the following actions (which are called 'passing a resolution') must also be approved by the shareholders:

- Change of company name
- Removing a director
- Changing the company's Articles of Association

To ensure that your company is compliant with regulations, you need to check that the information Companies House has about your company is correct every year. You do this by

---

8   Legislation.gov.uk, *Companies Act 2006*, www.legislation. gov.uk/ukpga/2006/46/contents, accessed April 2024

submitting an annual Confirmation Statement, which contains the following:

- Details of your registered office, directors, secretary and the address of where you keep the business records

- A statement of capital and shareholder information

- Your SIC code

- Your PSC register

There is further accounts reporting to HMRC that is required, but this will be covered in Chapter 6. Other annual compliance-related tasks will include renewing ICO membership (if required), IPSE membership and, of course, insurance.

Liability insurance

Whether you're a freelance web developer, a cybersecurity consultant, or an Artificial Intelligence (AI) wizard, it's easy for mistakes to happen in the digital realm, and the fallout (and associated cost) can be huge. As such, liability insurance is a critical component of risk management for IT contractors.

Professional indemnity insurance

While contracts should outline responsibilities and liabilities, having adequate insurance coverage adds an extra layer of protection. In most cases, this is non-negotiable, as IT contracts usually specify a minimum amount of PI insurance that is required to be in place. Think of PI as your digital safety net. It safeguards you from the financial fallout of any claims alleging professional negligence, errors or omissions in your services. Common examples in the IT field include buggy software causing operational chaos, or accidental data breaches. In all such cases, PI will cover legal costs and compensation payments.

Beyond these common examples, PI also offers protection against copyright infringement, defamation claims and inadvertent breaches of contract. As well as being a contractual requirement, PI will give you peace of mind that your business won't go under should you be involved in a legal dispute.

For IT contractors, the cost of PI is a worthwhile investment, usually just a fraction of the potential liabilities it guards against. As well as protecting your finances, it will also help you to establish trust with clients as it demonstrates your commitment to professionalism and risk management.

Public liability insurance

Away from the digital realm, there is another valuable insurance product, public liability insurance (PLI). It is common for PI and PLI to be included in the same policy.

PLI is your safety net against claims arising from injuries or property damage caused by your business activities to anyone except your employees (for this you need employers' liability insurance). It covers a wide range of scenarios, from broken windows to rogue chargers. It's easy for a harmless mishap to turn into a financial nightmare, and IT consulting often involves expensive equipment and offices. PLI will cover you for any such claims that may arise during a project.

Employer's liability insurance

If you have employees working for you, your business is legally required to have employer's liability insurance (ELI). There are some exclusions to this requirement, such as if you only hire family members.

ELI is like PLI, but covers accidents or injuries that involve specifically your staff. For example, a repetitive strain injury from years of data entry forces an employee to take early retirement – without ELI, your business could be lumped with huge potential legal costs and employee compensation payments. If you employ anyone, you cannot neglect this.

There are lots of options out there for insurance companies, and I would advise using a broker for PI and PLI. The Association of IPSE can help with insurance for jury duty (non-billable days payments) and IR35 legal protection and advice.

Annual compliance checklist

To make things easy, below is a comprehensive checklist of what you need to be checking, reporting and submitting annually to ensure your business remains legal and compliant.

Annual Reporting (Compliance)

- Confirmation Statement: Annual confirmation of information about and changes to your limited company.

- ICO (data protection) fees: Data protection registration fee.

- IPSE membership: Your insurance for jury duty and IR35 protection.

- Accounts: Annual accounts to be filed with Companies House.

- Company tax return: The statement of Corporation Tax due to HMRC.

- Self-Assessment tax return: The statement of Income Tax due from you personally (as a director) to HMRC.

- Insurance: Annual renewal of PI, PLI and ELI.

- Registered office: Annual renewal of your 'virtual office' address. Sometimes this is paid for alongside your company accounts.

Summary

In this chapter, we have covered the **annual renewals and reporting** you are responsible for keeping up to date with to ensure that your company remains compliant with laws and regulations. You should have systems in place to ensure reporting takes place at least annually and that you are meeting all requirements.

Companies House and HMRC must be provided with annual accounts and be updated with any changes to company directors and shareholders. It is essential that you have **professional insurances** in place.

To ensure you remain compliant and don't get any nasty surprises, complete my **compliance checklist** annually.

5
Accounting

I t's time now to turn to the subject of tax planning. It's essential that you pay the correct amount of tax. It sounds simple and it should be, but years and years of rules changes and additions have created a complex web of regulations that can make it difficult to understand the implications of your decisions and actions.

I am not a tax expert, but I can share some general information and guidelines that will give you a level of understanding and enable you to at least speak the same language as your tax adviser or accountant.

In essence, you need to record all transactions, consult an expert and listen to their recommendations, and record all your decisions and actions.

Record keeping

As a business, you must record all transactions affecting the company. In terms of income, that means invoices for sales, records of interest earned and other investment income that increases the value of the business, as well as statements for the company bank account. Outgoings or expenses decrease the value of business, and include travel costs, purchase of capital equipment such as laptops or computers and, of course, drawings from the business to your bank account.

If you do not keep accounting records, you can be fined £3,000 by HMRC or disqualified as a company director. You must keep accounting records for six years from the date of the last company financial year. In some cases, records must be kept longer than this – for example, if you purchase equipment that lasts longer than six years.

than 40% of your working time) there lasts or is likely to last more than 24 months. But if that work accounts for less than 40% of your total working time, then that location is considered temporary.

In practice, this means that if you have a customer location as a workplace (assuming you are on-site at least three days per week) for less than 24 months, you can claim for travel to and from that location, including any overnight stays and purchases of food during the duration of the temporary work. But if it exceeds that duration, those travel expenses are no longer allowable. You should take this into account when considering a renewal of a contract that would take the total contract to more than 24 months, as you may need to negotiate on day rate or location to account for the now non-allowable expenses.

Capital expenses

Capital expenses are equipment, machinery and business vehicles. There are various allowances for how much can be claimed for tax purposes. It is best to check with a tax professional before making any significant purchases. As an IT contractor, this is most likely to be your laptop. The business can and should

The act of recording these transactions is called book-keeping. At its most basic, it can be a spreadsheet detailing each transaction. In the case of an invoice, this would record the customer, the date of the payment, the total amount of the transaction, a record of VAT involved, a reference and a category, which would be sales. This transaction can then be reconciled against your bank account statements.

Records of expense transactions should contain similar details: where the transaction occurred (for example, the name of a hotel), when (a date), a reference, amount and any VAT paid. As with income, the expense transaction also needs to be given a category. In accounting, the category is important as, if recorded incorrectly, it can change the expense from an allowable business expense to a personal liability. We will go into more detail about expenses in the next section.

Sometimes, a spreadsheet will not be enough. If you have more than one customer, or many categories, record keeping and reporting will be clearer if you use accounting software, such as Xero or FreeAgent. This will incur a monthly charge but will enable more detailed reporting, which is useful for VAT and Corporation Tax returns, in addition to providing a clearer picture of the business. These are simple enough to use, given a little direction, and you can find book-keeping services if you prefer.

I strongly advise that you work with an accountant, especially when dealing with tax and expenses. Accountants can save you a lot of money, which will usually cover their costs.

Expenses

As a business, you will incur expenses. These might be as simple as paying for a new laptop or buying train tickets to get to a customer meeting. All such payments and purchases need to be recorded as expenses and assigned a category for accounts purposes, and associated with either a receipt or an invoice from the supplier.

Not all expenses are treated equally. There are three things you need to know about how to categorise an expense so that you are minimising your tax liability: is it allowable (wholly business related), is it a capital expense, or will it create a personal tax liability?

Allowable expenses

For allowable expenses, the business can off-set the expense against profit when calculating Corporation Tax. The general rule of thumb is that, as long as the purchase is a genuine business expense, it is 'allowable'. There are, fortunately, lots of guidelines about allowable and capital expenses, which are well known to accountants, and you can access them yourself on the gov.uk website or find the links in the Further Resources section. Generally, to be allowable:

- The expense must be wholly and exclusively for the purposes of the business

- The expense must be incurred in the current accounting period

- You must have proper records (eg receipts or invoices) to support your claim

There are some 'gotchas', though. For example, you can claim travel costs, if they are for business trips to temporary places of work. In contractor speak, we refer to the 24-month rule. A place of work ceases to be 'temporary' when continuous work (meaning more pay for this, and it will then be considered an asset of the business.

Personal tax liabilities

Commonly arising from 'benefits in kind', any personal tax liabilities are recorded on your P11D. A P11D is a form that must be submitted to HMRC by an employer annually for each member of staff (including directors) that receives certain benefits and expenses considered taxable by HMRC.

Anything can be purchased by the business, and care should be taken to ensure that you don't create an unwanted personal liability. If you use business funds for personal purposes, you may be taxed on that money as income. You can claim tax relief on entertainment of clients, suppliers or customers, but not on a meal out for yourself if you are not working at a customer site (temporary workplace).

An example of incurring a personal tax liability is that of Christmas or summer parties. At the time of publication, there is an allowance of £150 per member of staff, per year, that the company can spend and claim tax exemption on. It is important to note that this is an exemption, not an allowance. This means that if you have a Christmas party and either don't invite all the staff, or spend more than £150 per staff member, each staff member has a personal (tax) liability for that expenditure as if they had been paid it directly.

Paying yourself

Having a limited company gives you more choice in how you pay yourself. From a tax efficiency point of view, there are a couple of ways of drawing income from the company and into your personal bank account.

Any funds you personally withdraw from the company are subject to Income Tax and NI based on the amount of money withdrawn, and subject to certain tax-free thresholds. Dividends are taxed differently to salaries, and you have a specific dividend tax-free allowance and threshold (which has been diminishing in recent times).

More details and numbers are given in Chapter 6 HMRC And Reporting. The thresholds and percentages can change with government budgets, but the principles stay the same. I will provide illustrations as a guide, but you should always check the current rules and verify with your tax adviser / accountant.

As a contractor running your own business, you can decide what your salary is and what (if any) dividends you want to take. There is always the option of leaving money in the company for use later on – it could be a month later, or a year. If you don't need the money today, the best option might be to keep it in the company. It is up to you to get to grips with the rules and allowances and ensure you pay the correct amount of personal tax for the level of income you want to draw from your business.

As a company director, your pay will be split into four categories, each of which has its own rules and limitations. There are different thresholds / tiers for each category that will inform your decisions about how much money you want to assign to each category of personal drawings from your company. The four categories are:

- PAYE
- Dividends
- Pension
- Expenses

PAYE

As the director of the company, you are allowed – encouraged, even – to draw a salary. You'll want to pay yourself enough via PAYE to take advantage of your personal tax-free allowance, as specified by your tax code, but not too much to incur an unnecessary tax liability. If you pay yourself an amount under the NI ER threshold, there will be no NI liability at all. This may or may not be desirable, depending on whether you want to take advantage of the State Pension, statutory maternity/paternity pay and other benefits that require NI contributions. The rules change, though, if you employ staff in addition to yourself as director.

PAYE is the regular Income Tax all employees pay on their wages, and will create a payslip. The general advice for company directors is that you pay yourself a small regular amount to utilise your tax-free allowance but be aware of the NI contributions thresholds.

Dividends

The other main method of drawing money from the company is by declaring dividends. All companies can issue notice of dividends, where shareholders receive a share of the company's profits. This might happen once a quarter, or every six months with an interim dividend and a final dividend. As the owner and therefore a shareholder of your limited company, you are entitled to take dividends. Paying a dividend will create a dividend voucher.

As the dividend comes out of profits that have already been taxed (Corporation Tax), dividends are taxed at a lower rate. As the company director, you can decide how much the dividend should be. This decision should be based on the amount of profit made by the business and the level of personal tax that will be due on the drawings allocated to this category.

When deciding how much of the profits to withdraw from the company as dividends, you should consider leaving a contingency fund to be retained by the business in case of future costs or change in certain circumstances.

Pension

At some point, you will want to be able to stop work. If you build up a pension pot during your working life, this will give you a choice on when that is and more options on how to provide for yourself and your family in the future.

If you have paid yourself enough salary, your NI contributions will entitle you to the state pension. This will give you a basic level of income, but it might not be enough to maintain your desired lifestyle. Building a personal pension pot will help to bridge that gap and ensure you have a comfortable income for your retirement.

When someone asks me when they should start saving for a pension, I reply with 'yesterday'. Compounding interest rewards time. This means your money will grow not only as a result of your contributions, but also from the interest earned on those contributions, and then the interest earned on the interest. The earlier you start, the greater the potential for your contributions to grow significantly.

You also receive some tax relief on pension contributions, which are not taxed as income. Pension payments made direct from the company to help you prepare for retirement are an allowable expense and, as such, reduce your profit and therefore your Corporation Tax bill.

As of the 2024/25 tax year, there is an annual allowance of £60,000 that can be paid into a private pension.[9] Pension payments will be recorded in your accounts.

In terms of how much to assign to this category, the main question is: how much will you need to cover your lifestyle? A common rule of thumb is that your pension pot should be around 15–20 times your desired annual income. That means for an income of £50,000 you should aim for between £750,000 and £1,000,000. A financial adviser can assist you in creating a plan for retirement.

Expenses

Most out-of-pocket expenses (expenses that you have paid for with personal funds) should be claimed from the company and paid back to you. There is also the option of getting a company credit card and paying for expenses directly with that, saving the admin associated with expense claims.

---

9   Gov.uk, 'Tax on your private pension contributions', www. gov.uk/tax-on-your-private-pension/annual-allowance, accessed April 2024

As mentioned earlier, additional tax liability may be created through certain expenses that constitute benefits offered to employees; these are considered by HMRC to be 'benefits in kind', where the company has paid for something that you have personally benefited from. These will be recorded on a P11D and taxed according to your personal income.

Once you know the figures for each category, the appropriate paperwork is created and you can transfer the money out of your business bank account to yourself.

Closing the business

There is a fifth way of taking money out of the business, but that is for when you are finished trading and ready to close. Any funds left in the business at this time can be distributed at the close of the company. The exit plan should be part of any decision about whether to keep money in the business or not.

Money in the company at close is usually taxed at a lower rate than Income Tax, but you should check this before you decide – as I have stated many times now, tax rules change often.

Summary

Accountancy is a major and ongoing part of running any business, and IT contracting is no exception. The main bulk of your accounting practice will consist of **record keeping**, where you must keep detailed and up to date records of all of your sales (through invoices) and expenses, **categorising expenses** as you go. I recommend you use accounting software and an accountant to assist you with this.

The rest (if you are the only employee of your company) will relate to how you pay yourself and **categorise the drawings** you take from the company. Once money is received from customers, and a percentage set aside for tax, you can pay yourself from the profit. The funds that you intend to draw out of the company to pay yourself must be allocated into four piles:

1. Weekly / monthly **salary** paid to you via PAYE: This enables you to use your personal tax allowance and claim a state pension via NI contributions.

2. Monthly / quarterly **dividend**: This will be the rest of the income that you need now to live and look after yourself and your family.

3. **Pension contributions**: You want to retire at some point, right? This is money you draw from the company to give you an income in the future.

4. **Expenses**: You can and should reclaim any business expenses that you have paid for out of your own pocket.

This has been quite a technical chapter, but you should now understand the four categories of drawings that you can use to pay yourself, and how they affect your personal and company tax liability. This should guide you, and any advisers, in the decision-making process.

6
HMRC And Reporting

With over 500 different digital tax services, there are a lot of taxes that can
be reported on. Thankfully, as a contractor, we
need to focus on just a few:

- **VAT** – effectively a transition tax paid on
  expenses and charged on invoices.

- **PAYE** – a bunch of pay-as-you-earn taxes
  including Income Tax, NI (for employees
  and employers), and reporting on tax
  allowances, tax-free childcare and so on.

- **Corporation Tax** – paid on a limited
  company's profits.

- **Self-Assessment** – as a director, you must report on your total income and the correct level of tax due on those earnings.

VAT

VAT, as we've already discussed, is a tax levied on most goods and services sold in the UK. Businesses registered for VAT must collect it at the point of sale and then pay it to HMRC. The consumer ultimately bears the cost, reflected in the higher price of the product or service.

The UK currently has three main VAT rates:

- Standard rate: 20% – applies to most goods and services.

- Reduced rate: 5% – applies to specific items like energy-saving products and certain foods.

- Zero rate: 0% – applies to essential items like food, water and children's clothing.

As an IT contractor, if your turnover reaches the threshold, you will most likely need to be charging VAT. For example, if you charge £500 for some work, you should send your customer a bill for £500 + VAT = £600.

You would then keep this invoice as a record, both of income earned and VAT collected. As we discussed earlier, registered businesses must keep detailed records of their VAT transactions, including all sales, purchases and adjustments. These records are used to complete quarterly VAT returns, which are submitted electronically to HMRC. The balance between the VAT you have charged on sales, less the VAT you have paid on your own expenses, is what you will have to pay to HMRC.

The Flat Rate Scheme

You might be eligible for the Flat Rate Scheme if your *net* turnover is less than £150,000.[10]

The UK VAT Flat Rate Scheme is a simplified way for small businesses to deal with their VAT. Instead of keeping detailed records of the VAT you charge on sales and the VAT you pay on purchases, you simply apply a fixed percentage to your gross turnover and send that

---

10   Gov.uk, 'The VAT Flat Rate Scheme', www.gov.uk/vat-flat-rate-scheme, accessed April 2024

amount to HMRC annually as your VAT payment. For computer and IT consultants, the rate is currently 14.5% of your gross turnover.

While it means less paperwork, with only one annual return rather than four quarterly ones, it means you cannot claim VAT on most purchases. If you have low expenses, this isn't a problem, but if you are working away from home (temporary workplace) and paying for lots of hotels and restaurants then you are likely to pay more VAT than you would under the normal scheme.

PAYE

As an employee in the UK, you are taxed 'at source', via deductions made by employers before wages are paid. This is done via the PAYE, or Pay As You Earn scheme. Your payslip details your gross salary, then Income Tax, NI (paid by the employee and the employer), other deductions (pension contributions or student loan repayments, for example, and a total of what you will receive. As a business director, your company can and should pay you (partly) via the PAYE scheme, as we discussed earlier.

There are two main types of tax covered by this scheme: Income Tax and National Insurance.

Income Tax

This is a personal tax paid on your earnings. Everyone gets a Personal Allowance, which is the amount of income you can earn before paying any Income Tax. Currently, the standard Personal Allowance is £12,570.[11] Once you have reached the earnings threshold, you will begin to pay Income Tax on anything earned above this, at various rates. The current tax bands and rates are given in the table below (correct as of the 2024/25 tax year and subject to change with government budgets):

Tax Band	From	To	Income Tax/%	Dividend Tax/%
Personal Allowance	0	12,570	0	0
Basic Rate	12,571	50,270	20	8.75
Higher Rate	50,271	150,000	40	33.75
Additional Rate	150,001	+++	45	39.35

---

11   Gov.uk, 'Income tax rates', www.gov.uk/income-tax-rates, accessed April 2024

You will have a tax code, issued by HMRC, that will be used by payroll to calculate the amount of PAYE tax to be deducted from your gross salary. Your tax code is usually the full personal tax allowance, with the last number changed to an L. If you owe tax from a previous job/year, the code will be adjusted accordingly. Personal tax liabilities arising from P11D can also affect your tax code.

National Insurance

NI is a separate tax paid by both employers and employees, also through PAYE (or Self-Assessment if you are self-employed). Your NI contributions make you eligible for various state benefits, such as the Basic State Pension. For an employee, the amount is deducted from their gross salary and based on a percentage of earnings and specific income levels.

There are various classes of NI, we are going to discuss Class 1 (payable by employers and employees) here. The rules are slightly different for the self-employed. The NI employee (NI EE) contribution for most people (Category A, which accounts for the majority of employees) is currently 12% on earnings between £12,570 and £50,270. The company/employer then also pays NI based on the employees' earnings (NI ER). This is paid direct to HRMC by the employer and is considered a cost in addition to the gross pay. For Category A employees, the NI ER rate is currently 13.8% on earnings above £9,090.[12]

Allowances

The PAYE story doesn't end here. To reduce the tax burden, there are various tax allowances that you may be entitled to use.

- **Personal Allowance:** As summarised above, this is the amount of income you can personally earn tax-free. Some people, such as those over seventy-five or certain disability benefit recipients, may have a higher Personal Allowance.

- **Marriage Allowance:** You may be able to transfer 10% of your Personal Allowance to your spouse or civil partner if they

---

12 Gov.uk, 'National Insurance rates and categories', www. gov.uk/national-insurance-rates-letters, accessed April 2024

have lower earnings than you. This can reduce your tax bill.

- **Blind Person's Allowance:** If you're blind or severely sight-impaired, you may be eligible for an extra Personal Allowance.

- **Other allowances:** There are various other allowances and reliefs available for specific situations, such as charitable donations and pension contributions. You can find more information about these via your accountant or a tax expert.

Tax returns

The main form of reporting to HMRC is via tax returns. As a business owner, you will need to complete two types of tax return, with or without the help of an accountant: Corporation Tax and Self-Assessment. These are the two biggest tax bills you will have each year; let's look at each in turn.

Corporation Tax

Corporation Tax is a tax levied on the profits of limited companies and other incorporated businesses in the UK.[13] It's like Income Tax for individuals, but applied to companies instead.

Once the sales income has been paid into the bank account and your expenses – including director drawings of whatever amounts you decide – the business is left with its profit. It's this profit that is subject to Corporation Tax. The amount you need to pay will be calculated by your tax adviser along with your personal Income Tax (via Self-Assessment) once a year, and both are paid annually.

You can reduce your Corporation Tax liability by claiming allowable expenses. As explained earlier, allowable expenses are those that you incur while running your business, as well as pension contributions. An accountant will ensure that you are claiming for all the allowable expenses that you are entitled to, and not claiming for anything that isn't allowable.

Any expenses that the business has paid for but are not allowable do not count as a cost

---

13 Gov.uk, 'Corporation Tax', www.gov.uk/corporation-tax, accessed April 2024

for profit figure calculations for Corporation Tax purposes.

The Corporation Tax calculation looks complicated but can be simplified to this tiered table:

Corporation tax bands

Tax Band (£)	Tax on Band/%
0–49,999	19.0
50,000–249,999	26.5
250,000 +	25.0

The two rates of 19% and 26.5% means the actual rate between £0 and £250,000 is not more than 25%. We'll look at how much you'll need to budget for this in the next chapter. As with all tax information, these rates and bands are subject to change with government budgets. You must pay any Corporation Tax due no later than 9 months and 1 day after the end of their accounting period.

Corporation Tax can be pretty complex and it's not something you want to get wrong. It's always best to seek professional advice if you're not sure about your liabilities and I strongly recommend you use an accountant to prepare your Corporation Tax return.

Self-Assessment

When you have a limited company, as a company director, you will be registered for a Self-Assessment tax return. Self-Assessment refers to the system by which individuals report and pay their own Income Tax. This applies to those who receive income not automatically taxed through the PAYE system.

Directors of limited companies, people with multiple employments where tax isn't deducted at source from all and high earners under PAYE with additional income exceeding £100,000, all need to file a Self-Assessment tax return. As an IT contractor, multiple of these criteria could apply to you and you will almost certainly need to register for Self-Assessment.

Your Self-Assessment tax return is a yearly report that you are responsible for filing with HMRC, where you declare all your income, expenses and how much tax you owe. It will need to be submitted online (by you or your tax agent) by 31 January following the tax year to which it relates. For example, you would need to submit a return by 31 January 2025, for income earned in the 2023–2024 tax year.

The tax calculation includes all income from that tax year. This is why classifying your income via your accounting system is important, as earnings from each classification come with different rules. The different taxes you will calculate and pay via Self-Assessment include:

- Income Tax: As discussed, this is a progressive tax, meaning that you pay more tax as your income rises.

- Dividends: Dividend Tax is payable on all dividends you receive. This includes any publicly traded shares as well as your private shareholding in your company.

- Bank interest: With recently increased rates, this is now more relevant again. You have a separate personal allowance for bank interest; once this is reached, it is then taxed as income.

- Capital Gains Tax: If you sell assets (houses, cars) for more than you paid for them, you will have to pay Capital Gains Tax. The current rate is 20% for most assets and 32.5% if you fall into the higher rate tax band.

I have an accountant and tax adviser calculate and file my Self-Assessment tax return for me. It needs to be calculated and authorised by myself first, but with their help I can maximise allowances that generally only a tax professional will know about.

You are typically required to make advance payments ('payments on account') towards the next year's tax bill. These payments are typically due on 31 July and 31 January of the tax year. For example, for the current tax year (2024–2025), the first payment was due on 31 July 2024, and the second on 31 January 2025. The amount you need to pay is calculated based on your previous year's tax bill (split across two payments) and are essentially estimated prepayments towards your upcoming tax liability.

Summary

As an IT contractor, you will be responsible for paying your own taxes; these include **VAT, Income Tax, National Insurance** and **Corporation Tax**. It is important to plan for this so that you do not have any surprises at the end of the tax year. As the UK tax communications like to remind us, tax doesn't have to be taxing, but it does need to be planned and accounted for.

Being VAT registered will mean you are responsible for filing a **VAT return** every quarter (three months), or every year under the **Flat Rate Scheme**. As a limited company director, you are responsible for filing a **Corporation Tax** return once per year, to be paid 9 months and 1 day after the end of the accounting period. On a personal level, you will need to tell HRMC what your personal **Income Tax** and **National Insurance** liability is, via an annual **Self-Assessment**.

Claiming **allowable expenses** will reduce your overall tax liability. I strongly recommend that you use a tax adviser and/or accountant; they can file your Self-Assessment return for you along with the VAT and Corporation Tax returns, and ensure you make use of all allowances.

7

Budgeting And Cashflow Management

As the director and shareholder of a limited company, you have a responsibility to – and for – another legal entity: your company. Unlike your children, your company is under your complete control. Cash (as in currency) comes in via sales, flows through the business bank account, and some goes out again via expenses, leaving the difference as profit. You will need to manage these funds at all stages to ensure you can pay for expenses, pay yourself, meet your liabilities and make a profit, especially as some of these liabilities will have a hard deadline. The below image gives a rough idea of your main regular payments:

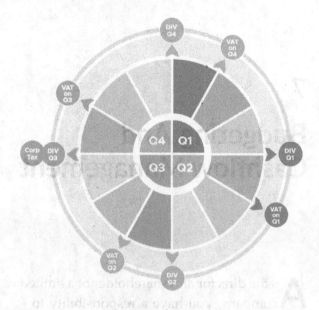

Financial management is a massive part of running a successful business in any field, and IT contracting is no exception. The number one reason for business failure is a lack of cash, and lack of cash is often due to poor financial management. This is important to address, as it will stop you from achieving your business goals.

Cashflow management

There is a rhythm to the cashflow of a business. There will be weekly or monthly invoice payments from your customers, and monthly, quarterly and annual payments to yourself and the government. You'll need to keep track of all funds coming in and out of your business, as well as dates that invoices and bills become due. The best way to do this is through monthly and quarterly reporting.

Monthly, the customer report will be useful to let you know if payment is late and needs chasing. Your monthly profit and loss report will show where your company stands.

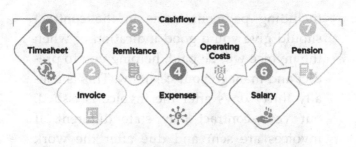

Monthly reporting

Quarterly reports include the VAT return, and again profit and loss, to determine if your planned level of dividend declaration is still affordable.

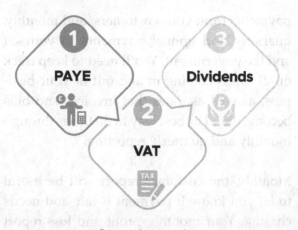

Quarterly reporting

Including payment terms in the contract should give you a good indication of when the invoice will be paid, helping you to predict and manage your cashflow. This is usually thirty days after the invoice is issued, but your contract may state different. If invoices are sent and due after the work has been commenced or completed, you are effectively allowing your client to pay you on credit.

Businesses are given credit ratings, just as individuals are. It is wise to run a credit check on your customer, before you sign any contract or renewal, so that you can be more informed about how solvent (having money in the bank to pay bills as they become due) they are. A credit check, using a platform such as Creditsafe, will show how long they take to pay an invoice, and give you an indication of how often they pay invoices late.

When raising your invoice, the customer will usually expect a timesheet that can be 'authorised'. This is a control that your customer will use to double check and authorise the invoice amount. Before payment is received, some customers will send a remittance note. This will detail the invoice reference, payable amount and when you can expect the funds to arrive in your bank account.

Once the payment is received in your business bank account, the payment should be recorded and the invoice is marked as paid on your accounting system. An allowance should be made for VAT and Corporation Tax should be calculated and set aside, and the rest contributes to the company profit.

The cashflow template below is a helpful way to allocate the money that flows into your business and will serve as a guide and budget, until you get the actual figures following a discussion with your adviser.

Invoice	12,000
Expenses, VAT and Corporation Tax	5,000
PAYE	900
Dividends	3,200
Pension	1,500
Contingency	1,400

Expenses

PAYE

Dividends

Pension

Contingency

VAT

Invoice

Cashflow template

Budgeting for tax

Tax will be the largest expense the business incurs and have the biggest impact on cashflow, and you must ensure that you have sufficient funds in the bank to cover it before you can think about taking any money out of your business. Fortunately, tax payments are easy to predict and plan for, as there will be a regular cycle of quarterly VAT payments and annual Corporation Tax to pay.

This section uses the rates and rules for 2024 but, as ever, bear in mind that tax rules change from year to year, so do check these regularly to ensure your budgeting is correct.

Budgeting for VAT

The VAT rate is set at 20%, so in terms of cash will account for approximately 16% of your gross sales transactions. That means that if your day rate is £500 and you bill an average of twenty days per month, the quarterly VAT figures will look like this:

£500 x 20 days	£10,000
VAT	£2,000
Gross monthly	£12,000
3 months per quarter	£36,000
Total VAT	£6,000

In the above example, the maximum amount of VAT that will be due to HRMC at the end of the quarter will be £6,000. Remember, as you also pay VAT on expenses that you reclaim in your VAT return, this will reduce the total figure owed. I always budget for the maximum amount as not all expenses are allowable for VAT. The VAT report run every quarter will tell you the exact amount due, and any surplus can be allocated or kept in the business.

Budgeting for Corporation Tax

This is a little trickier than VAT as the rates vary. As of 2023, there is no longer a single Corporation Tax rate and HMRC provide a Marginal Relief calculation for companies whose profits are lower than £250,000. This means that Corporation Tax ranges from 19% of profit (if your profits are below £50,000,) to 25% (if your profits are above £250,000).[14] As a rule of thumb, though, you know that it will be no more than 25% of your profit.

Assuming you have no expenses but pay yourself £10,000 per year via PAYE, your Corporation Tax calculation, depending on which bracket your profit falls into, would be as follows:

Without Marginal Relief

Sales	£260,000
PAYE	£10,000
Profit	£250,000
Tax @ 25%	£62,500
Or	
Tax @ 19% £50,000	£9,500
+ Tax @ 26.5% £200,000	£53,000
	= £62,500

---

14 Gov.uk, 'Corporation Tax rates and allowances', www. gov.uk/government/publications/rates-and-allowances-corporation-tax/rates-and-allowances-corporation-tax, accessed April 2024

With Marginal Relief

Sales	£110,000
PAYE	£10,000
Profit	£100,000
Tax @ 19% £50,000	£9,500
+ Tax @ 26.5% £50,000	£13,250
	= £22,750

To work out the likely Corporation Tax bill for your company and find out if you can claim Marginal Relief, you can use the calculator available from HRMC.

Before paying anything out to yourself and/or shareholders, your business should retain at least the Corporation Tax due and a quarter of VAT. In the examples above, this would be £23,100 or £68,500 for the higher-profit company. During your first year of trading, this total will be accrued over the course of the year.

Taking money out of the business

Before taking money out of the business, you must ensure that you will be able to pay for expenses that you incur during operations – insurance, capital purchases, professional fees, travel – out of the company bank account. My business has direct debits set up for my pension, my compliance adviser, accountant and indemnity insurance. Having these in place makes it easier to budget and calculate how much I can pay myself.

Then of course, taxes will also need to be paid. There will be VAT quarterly or annually and Corporation Tax annually. You may also have PAYE-related payments to make, depending on the amount you choose to pay yourself.

You will need to keep records for any money that you take out of the business. For PAYE salary, this will be in the form of a payslip. For dividends, it will be a dividend voucher. Your payslip might look a bit like this:

Payments		Deductions	
Monthly pay	£757.00	Tax	£0.00
Salary	£9,090	National Insurance	£0.00
Taxable gross pay	£757.00	Payment	£757.00

with a dividend slip like this:

DIVIDEND VOUCHER NO 1

An Interim Dividend of 3200 pence per share has been declared on the shares of the company for the period end. Payment has been transferred to you for the amount of such dividend due in respect of the shares registered in your name as follows:

On 100 Ordinary shares of £1 each

Dividend £3,200.00

This voucher should be carefully retained and will be accepted by the Inland Revenue as evidence of the dividend paid.

The way that you pay yourself and take money out of the company can also have an impact on cashflow, mainly in terms of where the tax liability falls (with the company or with you as an individual). In the simplified example above, you would attract tax pro-rata (assuming a Personal Allowance) as follows:

	Tax rate	Gross	Tax
Income Tax (in Personal Allowance)	@20%	£757.00	£0.00
Dividend Tax	@7.5%	£3,200.00	£218.21
Corporation Tax	@19%	£3,200.00	£608.00
Total tax liability		£3,957.00	£826.21

The overall tax liability includes personal tax (via Self-Assessment) paid on the dividend (with Dividend Tax calculated using leftover Personal Allowance) of £218.21 and company tax (Corporation Tax on the dividend) of £608, giving a total of £826.21.

The total cost to the business in this example is £4,565.00 (£757 salary payment + £3,200 dividend payment + £608 Corporation Tax due on the dividend); the net total amount received by you after all tax deductions is £3,738.79.

Now compare this to the total liability if you paid yourself only via payroll (assuming a tax code of 1257L):

Payments		Deductions	
Monthly pay	£4,103.28	Employers NI	£461.72
£49,239.36 Salary		Tax @20%	£611.16
Taxable gross pay	£4,565.00	National Insurance @12%	£366.69
		Payment	£3,125.43

In this example, Income Tax (of £611.16) and NI (£366.69) is paid through PAYE only; there is no SA tax bill. There is no Corporation Tax to pay as PAYE is an allowable expenditure.

The total cost to business is the same, at £4,565.00, but the net total received by you is less, at £3,125.43.

The difference in net received (£613.36) is close to the extra NI liability (£828.41) which, as a business owner, is not required by law, so you can decide which is best for you.

Building up capital

As an alternative to taking money out of the business, you can build up capital. If you're budgeting for your tax liability, this money will be growing in your bank account for some time before you need to pay it. The cash you are holding in your business is your known expenses that are yet to be paid and your contingency fund for when you take that planned (or perhaps unplanned) holiday, or something unexpected comes up, like jury duty.

In addition, month on month, the VAT you have collected grows. So does the Corporation Tax you are setting aside. If interest rates are favourable, you could put these funds to work in a high interest rate bank account. But remember: they are not yours to spend. They are earmarked for payment of your future liabilities.

Summary

The **cashflow starts with issuing an invoice** to your customer. This sets in motion a process that hopefully includes your customer paying your business an agreed amount by an agreed date and ends with you building up profit and capital in your business that you can then withdraw or invest.

To manage your cashflow and ensure that you always have the funds to pay yourself, meet your liabilities and perhaps build up some capital, you will need a **budget** and a plan for allocating the money that flows through your business. **Financial management** should be straightforward if you have systems in place, as payments (both incoming and out-going) will come in **regular cycles** with clear deadlines and so are relatively predictable. Tracking your expenses will give you an idea of how much you're spending and what your VAT bill is likely to be.

You can run **monthly, quarterly and annual reports** to give you a clear picture of how your business is managing its cashflow and can use tax calculators to get a rough idea of what your tax liabilities will be ahead of your tax adviser or accountant filing your returns.

PART THREE

PROFESSIONAL GROWTH AND THE FUTURE

In this final part of the book, I will describe the ongoing journey that is ahead of you in your quest for further contracts. It's not just about luck; I will explain how to be strategic in your search for the next contract by leveraging your experience and network as they grow.

We'll also discuss the 'end game' – what to do when you're ready to move on from your IT contracting business, which could involve training, investments and pension planning.

8
Building A Network

Networking is hugely important to contracting and has been part of business life for decades. I'm not referring here to our usual type of computer network: I am talking about your personal network of professional contacts. A wide network of other professionals, work colleagues and recruiters will help you stay in the know and mean that you get to hear about upcoming projects and contracts, ready for your next move.

Importance of networking

In the competitive IT contracting market, skills, trust and credibility are key to securing contracts. A professional network, forged through meaningful interactions, establishes you as a trusted professional. Having people who can vouch for your skills, reliability and work ethic can make all the difference when seeking your next contract.

Growing your network through relationships that are built through your delivery of projects demonstrates the quality of your work and communication skills. Over time, you will have a group of advocates who not only recommend you for opportunities but can also serve as references, giving you further credibility to potential clients.

Not all opportunities are advertised openly. Many lucrative contracts are circulated within professional circles before being publicly posted. These opportunities may arise through word-of-mouth referrals, networking, direct outreach or internal job postings within organisations. A well-established professional network means you can tap into this hidden job market, giving you access to opportunities that may not otherwise be available, or even visible, to you through traditional channels.

Through your network of professional contacts, you can also gain early insights into upcoming projects and the specific needs of potential clients. This foresight enables you to tailor your skills and position yourself more effectively as the go-to resource for that next project. Contractors who proactively seek out these opportunities can find themselves with higher-quality contracts with better compensation, benefits and alignment with career goals.

Mentorship is a particularly valuable aspect of a professional network. Seasoned professionals can offer guidance, share industry insights and provide advice to help you navigate the complexities and challenges of an IT contracting career. With their guidance, you can avoid common pitfalls and get a different perspective on problems you encounter, aiding your professional development.

Seek mentorship from experienced professionals in your niche. A mentor can provide guidance, share insights and help navigate your career. You can also consider offering mentorship to others, especially those who are early in their careers. Mentorship is a two-way street that fosters mutual growth.

Staying in touch with your peers will enable you to learn about other networking events and learning opportunities. This will give you the chance to meet new people and learn about new trends in the IT world. Meet-ups are common if you know where to look and can help you keep your skills on point.

Effective networking strategies

Building your network is not about collecting business cards (are those still a thing?) and connecting on social media. We're interested in quality over quantity. The best networks are created through meaningful relationships that contribute to both sides of the handshake.

It's not enough to just have a network, you need to stay visible within it to ensure you are front of mind when relevant opportunities arise in your network. Regular participation in industry events, conferences and online forums will enable you to showcase your expertise, stay informed, nurture your professional relationships and meet new people.

Attend workshops, panel discussions and networking sessions to connect with like-minded individuals who share your professional interests. As you engage in discussions, showcase your knowledge and contribute to the community, you will position yourself as a thought leader and expert in your field. This helps to build your profile as a contractor profile, which will not only attract potential clients, but also enhance your reputation within the IT industry, making you a sought-after candidate for contracts.

To elevate your profile further, offer value and share your expertise. Provide insights, tips and resources that showcase your knowledge, perhaps by writing articles or publishing blog posts. This not only establishes you as a valuable member of the community but also encourages others to engage with you.

Networking is about building relationships, not just collecting contacts. It's an ongoing process. Take the time to understand the people in your network, their professional interests and challenges. Interactions should be meaningful and you should follow up regularly to maintain a genuine connection. Building trust is crucial for a strong network, so stay in touch, share updates about your career and offer help when possible.

Marketing

Marketing yourself as an IT contractor requires you to showcase your skills, build a strong personal brand and leverage various platforms to reach potential clients.

What sets you apart from other IT contractors? Define your niche and specific areas in which you excel and that you enjoy working in. Your brand should not only reflect your technical skills but also convey your reliability, professionalism and problem-solving abilities. A USP that makes you stand out from the crowd is also a great thing to have.

In the digital era, to market yourself well you need to establish an online presence, starting with a well-crafted LinkedIn profile. Ensure your profile is complete and stays up to date, highlighting your skills, experience and notable projects. GitHub, Reddit and industry-specific forums will also enable you to create and build an online profile to demonstrate your expertise and showcase your projects. Join groups on these platforms to learn and share your insights. You should already have a website for your business, but consider adding a portfolio page to showcase your work, and think about any additional information that could provide more insight into your capabilities to potential clients.

Request endorsements and recommendations from colleagues, clients or supervisors. Display these endorsements and positive testimonials on your LinkedIn profile and website to build trust, credibility and to influence those considering your services.

You should also craft a compelling 'elevator pitch'. This is a couple of sentences that succinctly describe who you are, what you specialise in and the value you bring to clients. The pitch should be easily customisable depending on your audience and the context in which you are sharing it. For example:

'I am Neil Millard; I specialise in DevOps. I have delivered a solution that enabled the database administrators to upgrade their software with minimal downtime, by themselves.'

Ideally, this preparation will lead to a scenario where clients are seeking you out for their next project. Nothing helps to lift day rates like more than one customer competing for your time.

Early in my career, I put together a presentation that I delivered to Puppet Camp in London. The title was 'Scaling Immutable Servers in AWS with Puppet.' This was at a time before we could use serverless and had to manage EC2 instances with patching and all the security that comes with it.

The talk demonstrated my expertise in infrastructure configuration as well as my love of helicopters, through the demonstration and the gratuitous use of helicopter examples throughout the show. I met a great number of interesting people at the conference and I presented the talk again the following year.

In hindsight, it would have been nice to have got some photographs to go alongside the online slideshow and video. Instead, I recorded myself and created a YouTube video that is one of the most viewed on my channel.

Reusing assets like talks will enable you to reach a larger audience for your material and hard work. It also helps secure further opportunities to give talks or presentations, all of which will contribute to your marketing efforts.

Summary

In this chapter we learned about the importance of cultivating a robust and genuine **professional network** in a market where skills, trust and credibility are essential for securing contracts.

Networking isn't merely about collecting business cards and making virtual connections. It is essential to foster **meaningful relationships**. Effective networking involves active participation in industry events, engaged discussions and expanding your digital presence on online platforms like LinkedIn.

We also talked about the art of marketing yourself as a contractor, highlighting the need to define a niche and build a **personal brand** that extends beyond just your professional skills. Your **online presence** is critical, and at a minimum you should have a well-crafted LinkedIn profile and a professional website with your portfolio and endorsements. Crafting a compelling **elevator pitch** will offer a concise yet impactful introduction that will set you apart from the crowd.

Encounters and events – for example, talks at conferences – can have a ripple effect that can multiply your marketing efforts, and assets can be reused elsewhere online to get even more value out of your efforts.

9
Continuous Professional Development

The IT contracting market is always moving; projects come and go, tools are invented, adopted and abandoned. Skills are in demand one moment, and quickly replaced by the next big thing. Staying up to date requires continuous training and research.

Once upon a time, virtual machines and the internet did not exist. There will undoubtedly be countless tools and technologies that form a huge part of our work in the next decade or so that do not yet exist. We are in the business of information technology, after all, which is all about inventing and exploring.

Navigating technological change

In the world of IT, the only constant is change. Changes in tools, changes in technology – and none of it in change control. Seriously though, we love new tools, new ways of becoming more effective and efficient at delivering stuff. In the last twenty years there has been so much change. Moore's law has expanded computing power to the point where we can now hold the power of a 1990 supercomputer in our hand. Mobile apps are now expected. Websites are still holding their own, but for how long?

In your chosen field of expertise, there will be a prevalent technology, tool or methodology – in project management, it could be PRINCE2 or Agile; maybe it's Infrastructure as a Service, code, virtual machines, virtual networks; at the web frontend, it could be jQuery, React, Typescript. Whatever it is now, it can and will change.

Technology is always moving and evolving and, as the expert, you need to lead or be very close to the front. As a subject matter expert, you will find yourself needing to seek out the latest and greatest, given the opportunities to use it in your daily life, and perhaps even present it at your local user group or a national conference.

Keeping up with the new 'way' or toolset is the very essence of our role as IT contractors. Sharing the knowledge with our peers, learning from our mentors and teaching our engineers of tomorrow, is just the continuation of IT life.

As mentioned before, your network of professional contacts will assist you in the ongoing effort to keep your skills up to date. No new idea, technology or tool is created in isolation. Your peers, mentors and sometimes even publications will give you a heads up about what's coming next.

As IT contractors, we are on the pulse of new events, thoughts and technologies. Who knows, maybe you will be the next Marshal Hashimoto, or Linus Torvalds?

Industry trends and adaptation

As IT contractors we are always looking for and fixated with the next great new thing. Right now, it's AI. If you have been working in this industry, you'll know that AI has been around for years, decades even.

Recently, AI has been thrust into the public eye, with most of the focus on two areas: chatbots and generative AI. In particular there has been a lot of attention on AI images, driven mostly by the possibility of deep fakes. From a professional perspective, we know that AI is a powerful tool that can help us write code, simulate worlds and run data models to find patterns.

Did you pick up that data engineering would be a massive thing? Data Warehouse, Data Lake and the toolsets that make them accessible are still hot topics in the boardroom. The next logical step is Large Language Models (LLM), so that expensive call centre staff can be supported by chatbots.

Infrastructure as Code (IaC) and Infrastructure as a Service (IaaS) are still very much in vogue, but what's coming up next? Serverless has taken root, which is good for the cloud providers, as it is hard to migrate to your own data centre if you don't have any servers to migrate.

The costs of hardware are continuing to get cheaper, especially in storage, which is handy, as LLM and AI take up a lot of space. They also require high end CPUs (computer processing units) and GPUs (graphics processing units). The ability to rent an expensive GPU for the length of your analysis run provides a more cost-effective solution than buying the hardware – for now.

Is Blockchain still a thing? I hear of it from time to time, but there is no scandal loud enough to attract the attention of the press right now. But again, that may change.

Mentioning these technologies is going to age this chapter badly, which serves to illustrate my main point: don't rely on books to find the next industry trend. You should be following authors, researchers and thought leaders on social media and listen carefully to your peers about what is happening in the chief technical officers (CTO) suite, as success in IT contracting requires you to keep an eye on the future and an ear to the ground so that you can adapt quickly to changing client needs.

Certification and skill development

In my experience, the main barrier to acquiring a technical skill or learning how to use a new tool is related to the language surrounding it. Using shared terms and understanding what the components are called will enable further conversations and collaborations. If you have only learned a subject by tinkering with a home lab, or third hand via a YouTube video of someone else's, a short course may be of use to unite the theory with the jargon.

As an IT contractor, committing to lifelong learning is essential for staying abreast of industry trends, honing your skills and adapting to new challenges and opportunities. Continuous learning ensures that your skills and knowledge remain up to date with the latest advancements in technology and industry best practices. This is essential if you want to continue to deliver high-quality solutions and remain competitive in the marketplace.

Learning new technologies, programming languages, or methodologies broadens your skill set and enhances your versatility as an IT professional. By diversifying your expertise, you will increase your value to clients and open up opportunities for new types of projects.

Acquiring new certifications, mastering in-demand skills and staying up to date with industry trends can pave the way for career advancement and higher-paying opportunities. In addition, some technology providers and companies will require that you complete specific training and certification for partnership.

Pursuing industry-recognised certifications demonstrates your expertise and proficiency in specific technologies or domains. Whether it's certifications in cloud computing (eg AWS, Azure), cybersecurity (eg CISSP, CEH), project management (eg PMP, PRINCE2), or agile methodologies (eg Scrum, Kanban), certifications can enhance your credibility and marketability as an IT contractor.

Understanding the industry trends should help guide you in deciding which certifications and training courses to pursue, to ensure you get the best return on your investment. Obtaining a certification can involve a significant commitment in terms of time and money, so you need to be confident that it will pay dividends.

In-person training can be a particularly expensive option, but as IT professionals we are perfectly placed to explore online learning platforms (such as Coursera, Udemy, or Pluralsight), which offer a wide range of courses and tutorials on IT topics spanning programming, cloud computing, cybersecurity, data science and more. Online platforms provide flexible, self-paced learning opportunities accessible from anywhere with an internet connection.

You can stay informed about the latest trends, best practices and innovations in IT in general by reading technical books, whitepapers, research papers and industry publications. You could also subscribe to relevant magazines, journals or online forums to access valuable insights and thought leadership from experts in your field.

Learning should always be put into action to properly embed the knowledge gained. Apply your learning in real-world scenarios by working on hands-on projects or side projects outside of your regular contracts.

Building software prototypes, contributing to open-source projects, or experimenting with new technologies in a sandbox environment allows you to apply and reinforce your skills while gaining practical experience without the risk of trial and error on a client project.

In your pursuit of continuous learning, you should aim to strike a balance between depth and breadth in your skill development. Deepen your expertise in core areas of specialisation while also exploring adjacent technologies or domains that complement your skill set. This balanced approach will enable you to remain agile and adaptable while capitalising on your strengths.

Balancing professional and personal growth

If you are charging your client a day rate, when do you find the time to train and learn? As a contractor, you might have spare funds that you can use for training to advance your professional development. A bonus is that these are allowable expenses for tax purposes. You can see it as a business expense, and take time off to train, or you can use your spare time. As a contractor and business owner, it is up to you to find a balance that gives you enough time to work on yourself and enjoy your hobbies.

To give you a personal example, I always wanted to fly helicopters. I was earning a decent day rate, so I felt it would be better to take half a day during the week to follow my dream of becoming a pilot. This left me with weekends to spend with my family.

I love learning and I love computing. I've found that I can easily spend too much time learning, experimenting and exploring a new subject, tool or skill and totally forget about anything else. I find that having a goal helps me to stay on course, and making diary appointments with myself for dedicated time where I can focus on learning, enables me to schedule my time appropriately, without becoming distracted by the next shiny toolset.

Personal growth is also important, and perhaps you want to work on more general skills, such as public speaking or writing. These can be used to enable great presentations and speaking engagements. It might be difficult to justify these as business expenses though, as they are not directly related to the services you deliver.

Summary

One of the key attributes of successful IT contractors is their ability to embrace change. Technology is constantly evolving, and as an IT contractor, you must **stay adaptable** and open to **new opportunities**. Whether it's mastering emerging technologies, adapting to different project requirements, or adjusting to shifts in the market, being flexible and proactive will set you apart and ensure you continue to win contracts.

In the fast-paced world of IT, the learning never stops. As you progress in your contracting career, **invest in continuous learning** to **stay relevant** and **competitive**. Whether it's through certifications, workshops, online courses or hands-on experience, prioritise expanding your skill set and staying abreast of industry trends. This not only enhances your marketability but also opens doors to new and exciting opportunities.

10
Advanced Contracting Skills

In this chapter we will discover how to get more from your contracts. We'll look first at the importance of client expectations to the smooth running of contracts, from beginning to end. This is aided by having a great relationship with the customer and excellent communication. The customer here might be one of many stakeholders in the project, or the team you are working with.

We'll also talk about the need to have a way to share progress and receive feedback. Communication is a two-way street and understanding the relationship and project from both perspective aids in mutual understanding.

Once you have a handle on client expectations and how to best communicate, you will be in a position to handle multiple contracts at once. This will require excellent organisational skills.

Client expectations

If one good thing came out of the global pandemic, it was (finally) the widespread adoption of working from home. The technology has existed for ages. Collaboration tools are readily available, it just took a little push to convince management that there was no impact on productivity.

Now that offices are becoming noisy with people again, there is a desire among some of my clients to return to office-based work full-time. This makes sense if the whole team you are working with return, but if even one person is remote, all the remote tools will still need to be utilised, making physical proximity irrelevant.

You will need to understand how hybrid working will be integrated into your service delivery, and this should be a clause in the contract – mostly as it affects the location from which you are expected to deliver services and can complicate expenses, as discussed earlier.

Most customers I work with want to reduce their office overheads, and so do not wish to pay for contractors to take up this expensive space, and meeting rooms are still at a premium.

The expectations of the client should be known before you commence the contract – as well as location, this includes project scope, deliverables, timetables and budget constraints – but as I keep saying, change is constant. The client will also have expectations around the level of service you will deliver. This will be determined by the standard you set with your work and will need to be managed should any hiccups occur. Keeping clear and open communication will ensure that the customer's expectations remain in line with what you can provide. Communication skills are key to keeping customers happy and well informed of both great and not so great delivery. Regular progress updates, status reports and milestone reviews can help ensure that everyone stays on the same page throughout the project.

Managing conflict

Despite your best efforts, conflicts may arise during the course of a project. It's essential to address these constructively and proactively. Take the time to listen to your client's concerns and perspectives without interrupting. Understanding their point of view is the first step towards finding a resolution. Clarify any misconceptions and ensure that you have a shared understanding of the situation.

Focus on finding practical solutions that address the root cause of the conflict and try to seek common ground – you may have to compromise. Finding a solution that meets everyone's needs can help preserve the client relationship and maintain project momentum. Once a resolution is reached, be sure to document this and distribute a written copy for everybody's records.

Ultimately, the renewal of your contract will be largely determined by the client's opinion of you and the service you have delivered. Keeping them informed and meeting (or exceeding) their expectations will put you in a good position to renew the contract, should you wish to.

Stakeholder communication and relationships

Stakeholders play a vital role in the success of a project, and good communication is essential for ensuring that their needs and expectations are met. Whether it's clients, project sponsors, team members or end-users, clear and concise communication leads to collaboration, alignment and trust.

When you start a new contract, create a communication plan that outlines how, when and what information will be shared with stakeholders throughout the project lifecycle. There are various communication channels you might use, including email, meetings, status reports, project management tools (eg Slack) and collaboration platforms (eg Trello) to ensure that stakeholders receive timely updates and remain engaged with you on the project.

It is important to choose the tools that the stakeholders are most likely to use. Make it easy for them to communicate with you. Not all stakeholders will have the same level of technical expertise or interest in project details. Tailor your communication approach and messaging according to the audience to ensure clarity and relevance. Use layman's terms when discussing technical concepts with non-technical people and always provide enough context to ensure they will understand what you're telling them. Depending on your role, your contact may be limited to collaborating on communications or writing them wholly with input from other team members.

It's important to provide your client with regular progress updates. Personally, I practise Scrum and Kanban, which creates a daily and weekly cadence. I find that this consistency in communication helps manage expectations and build confidence.

Effective communication is a two-way street. Encourage stakeholders to voice their opinions, ask questions, provide feedback, and input and express any concerns openly. Are they enjoying the reports? Do they prefer a different format? Show that you're interested in open dialogue and collaboration so that all stakeholders feel valued and empowered to contribute to the project's success.

Speaking of success, it's crucial to recognise and celebrate project milestones and achievements with stakeholders and the team. Whether it's reaching a significant development milestone, delivering a successful product demo, or achieving project objectives ahead of schedule, sharing successes reinforces stakeholders' confidence in your abilities as a contractor and fosters a positive project.

Managing multiple contracts

In the next chapter I will introduce you to consulting, but that may still be a little way into your future. Somewhere in between contracting and building a consultancy lies the task of managing multiple contracts. When you follow the advice in this book and build systems and processes that facilitate success as an IT contractor, you may find yourself in the position where your skills are needed by more than one customer at a time. Being in demand is a wonderful thing and puts you in the great position of having more choice.

If the required time commitment allows, you can split your time between multiple contracts or projects simultaneously. It will be a test of your time management and organisation skills to ensure that all commitments are met, and you are delivering the best service in line with your reputation.

I have found that to start with, limiting yourself to just one project per day and focusing on that project for a specific day each week, keeps things simple. It reduces context switching and provides a clear break between projects.

You must ensure that you allocate enough time for each contract based on the expectations of the client and your own bandwidth, so avoid overcommitting and be realistic about what you can get done.

Your comms skills remain important, as you will need to keep open lines of communication with the clients for all contracts. You must be sure that they understand your availability and the time allocated to them and be quick to quash any conflicts and inform of delays. Managing the expectations of multiple clients at once is a test of your skills but will ensure that you and all of your clients get what you need.

If the time commitment required is more than you have available, the customer may be happy with you providing a substitute for one of the contracts. In addition, delegating and outsourcing routine or administrative tasks to trusted individuals can free up your time for the high-value tasks. These people can be employed directly by your company or sub-contracted.

You can use technology to stay organised, with separate repositories for documentation, code and other relevant information for each contract. This ensures easy access to crucial resources and facilitates collaboration with clients and team members.

With careful planning, organisation, and communication, you can maximise your productivity and manage multiple contracts while maintaining a high standard of work and client satisfaction.

Summary

Managing customer expectations involves understanding and aligning with client needs and desires at all stages of a project. Maintaining clarity on project scope, deliverables, timescales and budget constraints will ensure client satisfaction and facilitate successful project outcomes.

Effective communication and **strong client relationships** are critical to a project's success. Effective communication involves clear and concise exchange of information between you and all stakeholders. Building trust through regular updates, open dialogue and tailoring messages to the audience will help to build strong relationships.

Managing multiple contracts and/or clients at once may involve some juggling to, but with **great time management**, realistic allocation of time and resources and clear communication you can serve all your clients and avoid conflicts. To free up time, delegate routine tasks so you can focus on maintaining high standards of work.

11
Future Planning And Transitions

You can have a long and fruitful career as an IT contractor, but what follows? In this chapter I will share my experiences and some tips for what comes next. You could branch out into IT consulting, which gives a slightly different view of the market and more varied engagement with projects. This option leverages your experience as a contractor to offer broader advice and expertise to clients, potentially requiring less hands-on work. It allows you to take more of a leadership or an architect role, depending on your skills and experience, with a few different directions you can go in.

Or perhaps you wish to exit contracting altogether and either transition to a permanent role or retire from the day-to-day entirely. In this case, having an exit strategy will help you make this transition smoothly from both a financial perspective and a personal fulfilment one.

Speaking of finances – we'll then go on to talk a bit about financial planning. This is a subject that I have touched upon throughout the book and it feels fitting that it rounds off the chapter. Right now we need money to fund our lifestyle, maybe this number will change in the future – it could go up or down – for now, financial planning will give us choices.

Consulting

Transitioning to a consultancy role might be the next logical step in your IT contracting career. If you decide it's for you, this path will require a change in mindset. You will need to adopt a strategic, long-term perspective that focuses on creating value, solving problems, and building a sustainable business that exists and endures beyond individual contracts or engagements.

Consulting is an evolutionary progression from contracting. The first step, which you may have already taken, is to find a niche. Reflect on your skills, expertise, and industry knowledge to identify a niche area where you can offer specialised consulting services. Whether it's cloud migration, cybersecurity, data analytics or software development methodologies, carving out a niche will help you to differentiate yourself in the market and attract clients seeking specialised expertise.

You will need to assess and mitigate the risks associated with moving into consulting and diversify your revenue streams, build and keep a financial buffer and create contingency plans to help you navigate uncertainties and unforeseen challenges.

Staying agile and adaptable in your approach will enable a smoother transition from contracting to consultancy. Continuously seek feedback, learn from failures and iterate on your strategies to stay ahead of market trends and jump on emerging opportunities. Surround yourself with mentors, advisers and peers who can offer guidance, support and insights on your journey. Learn from their experiences, leverage their networks and get their advice on strategic decisions and problems you face along the way. Despite the made-man image portrayed by some, consulting is a team sport.

Consulting requires a better understanding of sales and what the client wants, and you will need to clearly articulate the value proposition you offer. What unique insights, solutions or outcomes can you deliver to clients? Highlight your track record of success, industry credentials and proven methodologies to demonstrate your credibility and expertise.

As a consultant, you should develop frameworks, methodologies and/or service offerings that provide structure and clarity to your consulting engagements. This may include assessment tools, implementation frameworks, best practice guides, or proprietary methodologies that streamline your consulting process and deliver consistent results for clients. The 3Cs framework I share in this book is a great example of this.

On an ongoing basis you must invest in building your personal brand as a consultant by establishing thought leadership, sharing insights and showcasing your expertise through content creation, public speaking and industry events. To widen your reach, leverage online platforms like LinkedIn, professional blogs or industry forums; this will amplify your visibility and put you in front of potential clients.

Networking is just as important in consultancy as it is in contracting. You must cultivate relationships with potential clients, industry peers and referral sources, and always be on the lookout for people who may benefit from your consulting services. Again, industry conferences, networking events and trade shows provide great opportunities to connect with decision-makers and influencers in your target market. When networking, be sure to offer value through educational content, consultations or workshops to show off your expertise and build trust with prospective clients.

Consulting can allow you to enjoy greater freedom and autonomy in your career, with more flexibility in choosing your projects, clients and work schedule, allowing you to be more selective and align your work with your personal interests, values and long-term goals. This means you can pursue a healthy work–life balance with time for both professional and personal pursuits. But you will need to set boundaries, prioritise self-care and allocate time for activities that support your wellbeing and fulfilment outside of work, to prevent consultancy from becoming all-consuming.

These benefits come with increased responsibility and accountability and, as a consultant, you must be prepared to take ownership of your decisions, actions and their outcomes. Your priority must always be delivering value to your clients.

Entrepreneurship

There is another way out of contracting that isn't consultancy, employment or retirement. It's a lot of hard work and lots of fun. Entrepreneurship offers another avenue for IT contractors looking to transition out of contracting. If you have a passion for innovation, autonomy and building your own enterprise, consider starting your own IT business. We've talked about creating a consultancy firm – entrepreneurship is the next level.

If you think you could be an entrepreneur, draw upon your contracting experience to identify market opportunities, niche specialties, or pain points that you can address with your products or services. Then you'll need to develop a business plan, secure funding if necessary and leverage your network and industry expertise to launch and grow your venture successfully.

It's going to take hard work, so recognise when you need assistance and don't hesitate to delegate tasks or outsource projects that fall outside your expertise or capacity. That might mean hiring subcontractors, collaborating with colleagues or making use of automation tools and services. Delegating will free up your time and mental bandwidth to spend on the parts of the business that matter most to you and to focus on your core strengths and priorities; this will make you more efficient and effective.

Entrepreneurship can be highly rewarding, but there is a risk of burning out. You'll need to have a strong team around you and take time to prioritise your own health and wellbeing as well as that of your business. Make self-care a non-negotiable part of your daily routine, with dedicated time each day to spend on things that feed your body, mind and soul and bring you joy and fulfilment outside of work. That could be exercise, meditation, hobbies, spending time outdoors, with family or simply relaxing and unwinding. It's important that you get enough sleep, eat well and stay hydrated to keep up your energy levels and stay healthy.

Don't let entrepreneurship run away with you – make time to regularly reflect on your work–life balance and re-evaluate your priorities, habits and commitments to ensure you don't compromise your overall wellbeing and fulfilment. Take stock of what's working well and where things could improve and make changes if you need to. Be proactive in identifying potential sources of stress and take steps to address them before things escalate. By cultivating self-awareness and mindfulness, you can maintain a healthy balance between your work and personal life and allow entrepreneurship to bring you greater satisfaction and fulfilment in both domains.

Financial planning

Financial planning as an IT contractor is not just about budgeting for the year, but also preparing for the medium and long-term future. Creating a detailed personal budget is fundamental to managing your finances as an IT contractor. Since your income may fluctuate from month to month, it's essential to track your expenses meticulously and allocate funds for both necessities and discretionary spending. Consider using budgeting tools or apps to monitor your cashflow and identify areas where you can optimise spending. I use YNAB (You Need a Budget) for my personal finances, and FloatApp for financial forecasting based on my Xero accounting data.

Building a personal emergency fund is crucial for weathering unforeseen expenses or periods of low income. You should aim to set aside enough savings to cover at least three to six months' worth of living expenses. This buffer provides financial security and peace of mind, allowing you to navigate challenges such as unexpected medical bills, car repairs or temporary gaps in contracts, without undue stress.

Once you have an emergency fund sorted and you're paying regularly into your retirement account (more on this in the next section), you could then use your accumulated capital to pursue investment opportunities to grow your wealth over the long term. For example, you might invest in businesses, property or other ventures, generating passive income. Depending on your risk tolerance and financial goals, you can consider investing in stocks, bonds, exchange-traded funds (ETFs), real estate or other asset classes.

You should aim to diversify your investment portfolio to minimise risk and maximise potential returns, and periodically review and rebalance your investments to align with changing market conditions and personal objectives. All forms of investment come with risks attached, so seek proper advice and invest in areas you know about so that you are better informed about the potential risks. If you'd prefer not to invest, spare cash can be kept in the business and invested in high-yield cash savings.

Navigating the complexities of financial planning as an IT contractor can be daunting, so it's a good idea to seek professional guidance. Consider working with a certified financial planner or adviser who specialises in serving contractors and understands the unique challenges and opportunities of your profession. They can help you to develop a comprehensive financial plan tailored to your individual circumstances, needs and goals, which you should regularly review and adjust as needed so that you stay on track to achieving financial success.

Exiting contracting

At some point in the future, you may decide that you're done with contracting and you're ready to move on or retire. In which case, you will need to close your company and think about how you're going to fund your life.

Retirement

Planning for retirement is critical for long-term financial security. Earlier in the book we talked about pensions. Saving into a pension is wise as funds will be required for your future comfort when you take the decision to stop working. You should contribute regularly to your pension to build a nest egg for your future and take advantage of tax benefits and compound growth over time. Not only is it tax efficient, but it will also give you peace of mind about your eventual exit from the work force.

When you retire, the company will cease trading and you must close it down. This is an easy process: you can end any compliance reporting, unregister for VAT and wind up the company. Any cash in the company can be withdrawn as a lump sum upon business bank account closure. If you work with an accountant, they will help you close the account and withdraw funds in the most tax efficient way for you.

Employment

What if you are not ready to retire just yet, but want to leave contracting behind and transition to a permanent role, perhaps with a client or a different organisation? You might decide that this better aligns with your career objectives. For example, if you find yourself coveting the stability, benefits and career advancement opportunities associated with permanent employment, you might be happier in a full-time position. If so, network with your clients, colleagues and industry contacts to uncover potential job openings and leverage your contracting experience and skills to position yourself as an attractive candidate.

If you exit contracting for a job rather than to retire, you may want to keep your company open so that you have the option to return in the future. In this case, rather than cease trading, you can file dormant accounts. There will be some fees associated with keeping your company open, but these are quite small. You will need to keep your insurances in place, particularly PI, as you could face a claim associated with previous contracts for up to six years. If you are sure you don't intend to return to contracting, closing the company might be the correct choice for you.

Summary

While you must of course focus on the present, it's just as important to plan for the future, including your eventual **exit from contracting**. This exit can take many forms – branching out into contracting, becoming an entrepreneur, moving on to a permanent role or retiring. Whatever you choose, having an exit strategy will provide clarity and direction for your career journey. Keep your skills up to date, nurture your network and stay open to new opportunities that support your long-term goals.

By entering the world of **consulting**, IT contractors can leverage their expertise and creativity to create new opportunities for growth and impact. With strategic planning, continuous learning and a proactive mindset, contractors can successfully move into consulting and thrive in the dynamic IT industry. If you have a particularly entrepreneurial spirit, you might take this one step further and **start your own IT business as an entrepreneur**.

As an IT contractor, **managing your finances** is crucial for long-term success and security, whatever career path you pursue. Creating a robust financial plan that accounts for variable income, taxes, insurance and retirement savings is essential. Consider working with a financial adviser who understands the unique challenges and opportunities of contracting to develop a personalised plan that will help you achieve your goals and aspirations. Good financial planning is crucial to ensuring you can **retire from contracting** at a time of your choosing and continue to live a comfortable life.

Conclusion

What have we learned throughout this exploration of IT contracting? The aim of this book was to provide a comprehensive understanding of the foundational requirements to successfully start and run a contracting business in the form of a limited company.

We started at the beginning, covering the critical aspects of company formation and worked through all the operational know-how necessary for efficient and compliant management of an IT contracting business. At the core of any contracting venture lies the 3Cs: Contract, Cashflow and Compliance. These three pillars form the backbone of a thriving and sustainable contracting practice.

Contract

We got into the nitty gritty of contracts, the lifeblood of any contracting engagement. A well-structured contract, as well as outlining the scope of work and deliverables, also protects you from potential disputes and misunderstandings. We looked at the essential elements of a contract required for a smooth and mutually beneficial collaboration between the contractor and the client. We also addressed the critical issue of IR35, providing insights that will help you navigate this important regulatory framework and mitigate any risks.

Cashflow

Effective cashflow management is the lifeblood of any business, and IT contracting is no exception. We looked at how to streamline your systems to ensure a steady inflow of cash into the business while keeping an eye on outgoings to ensure you can support your desired lifestyle and meet your liabilities. Understanding the timing and flow of funds is crucial so that you can make informed decisions about when to draw income from the company and how. By mastering cashflow management, you can achieve financial stability and avoid the pitfalls of irregular income streams.

Compliance

The third pillar, compliance, encompasses the regular recording and reporting obligations that must be fulfilled to maintain legal and regulatory conformity. We covered the monthly, quarterly and yearly reporting obligations required to stay compliant, including record keeping, tax planning, VAT calculations and personal liability considerations. Compliance can seem daunting at first, but by implementing the proper systems and processes, you can navigate these obligations with ease, minimising risks and avoiding costly penalties.

Finally, we explored the options available to contractors when it comes time to exit the contracting world. Whether you envision moving into consulting, pursuing investment opportunities or embracing retirement, having a well-thought-out exit strategy is essential for a smooth transition and the preservation of your hard-earned financial gains.

A final note: be sure to prioritise maintaining a healthy work–life balance throughout your contracting career. While the flexibility and autonomy of contracting is liberating, it can be easy to fall into the trap of overworking. Remember to schedule downtime, pursue your hobbies and interests outside of work and nurture relationships with family and friends. This will not only enhance and protect your wellbeing but also drive creativity, productivity and overall job satisfaction. Don't just survive, thrive!

The 3Cs to Become a Confident Contractor course

Ready to take your contracting business to the next level? This book has provided a solid foundation, but building a truly successful contracting business and maintaining it over the long term requires going even deeper. My '3Cs to Become a Confident Contractor' course offers an even more detailed exploration of the key areas covered in the book, along with the personalised support you need to thrive.

Here's what sets the '3Cs to Become a Confident Contractor' course apart:

- **Deep dives and actionable strategies:** We'll delve even further into crafting contracts, mastering cashflow management and navigating complex compliance requirements. You'll walk away with actionable strategies to implement immediately.

- **One-on-one support:** You won't be on your own. I'll be there to answer your questions, address your unique challenges and guide you through any roadblocks you encounter.

- **Proven success strategies:** Learn from my experiences and those of successful contractors who have benefited from the course, for even more valuable insights.

- **Community and networking:** Connect with other ambitious contractors, share knowledge and build a support network that will fuel your growth.

The '3Cs to Become a Confident Contractor' course empowers you to achieve your contracting goals with confidence. It's the roadmap you need to build a sustainable, profitable business and secure your financial future.

Find out more at www.confident-contractor. co.uk / a-brief-overview-of-the-3cs-course

Resources

Companies House, SIC classifications: www.gov.uk/government/ publications/standard-industrial -classification-of-economic-activities-sic

Creditsafe, For customer credit checks: www.creditsafe.com

Gov.uk, Allowable expenses: www.gov. uk/expenses-if-youre-self-employed and www.gov.uk/corporation-tax-rates/ allowances-and-reliefs

Gov.uk, Check employment for tax (CEST) tool: www.gov.uk/guidance/ check-employment-status-for-tax

Gov.uk, Company name checker: https://find-and-update.company-information.service.gov.uk/company-name-availability

HMRC, Corporation Tax and Marginal Relief calculator: www.tax.service.gov.uk/marginal-relief-calculator

IPSE (Association of Independent Professionals and the Self-Employed): www.ipse.co.uk

Acknowledgements

I would like to thank Nick Orme for his graphic expertise, which you can see in the illustrations for this book.

This book has also had invaluable input from my beta readers, Marti Floyd, Dominic Spinks and Gerald Benischke.

The Author

 Neil Millard is widely recognised for his technical expertise in leveraging infrastructure, CI/CD methodologies and Python APIs to create efficient workflows throughout the software delivery process. Neil is the author of *Who Moved My Servers?*, a book about key DevOps principles.

Recently, Neil won the prestigious DevOps Professional of the Year award at Computing's DevOps Excellence awards in London. As a DevOps expert at HMRC, the UK's tax authority, he specialises in cloud deployment leveraging tools like Terraform, AWS, Lambda functions (Python and others), ECS and AppMesh. His day-to-day work revolves around cloud deployments, automation, scaling and ensuring data security.

When he's not immersed in the world of DevOps, Neil pursues his passion for aviation by flying helicopters. He's also an avid scuba diver. Based in Bristol, UK, Neil offers DevOps consultancy services and mentors aspiring professionals on transitioning their careers to contracting in the UK DevOps market. He lives with his wife, Wendy.

If you're ready to join the programme and unlock your full potential, connect with Neil:

in https://www.linkedin.com/in/neilmillard/

✖ https://twitter.com/neil_millard

🌐 https://neilmillard.com

🌐 www.confident-contractor.co.uk